Mastering Your PhD

Patricia Gosling · Bart Noordam

Mastering Your PhD

Survival and Success in the Doctoral Years and Beyond

Third Edition

Springer

Patricia Gosling
Zurich, Switzerland

Bart Noordam
ASML (Netherlands)
Amsterdam, The Netherlands

ISBN 978-3-031-11416-8   ISBN 978-3-031-11417-5   (eBook)
https://doi.org/10.1007/978-3-031-11417-5

This Springer imprint is published by the registered company Springer Nature Switzerland AG
The registered company address is: Gewerbestrasse 11, 6330 Cham, Switzerland

Preface

Why Read a Book About Getting a PhD?

PhD students and their supervisors tend to focus only on the content of the research that leads to the doctoral thesis. All other issues are often taken for granted: how to organise your work, give a presentation, work in a team, cope with your supervisor, and manage your time effectively. When asked, former PhD students typically claim that the general experience of being a graduate student, which includes learning how to solve complex problems and work well with others, was of greater value to their careers than the actual content of their thesis. The goal of this book is to apply the communication and organisational skills found to be effective outside the realm of academia to the world of PhD students, and to encourage them to master the non-scientific aspects of getting a PhD. Hopefully, the suggestions and advice included here will help graduate students get the most out of (and fully enjoy!) their PhD years, in addition to offering some much-needed support as they prepare for their post-PhD careers.

Sink or Swim

After hearing about this project, a professor in the UK had this to say: 'This book should not be published. Obtaining your PhD is like swimming across a big lake. Some students cannot swim, so they'll sink. That's the way the academic system selects those who will win. By providing students with a book on how to swim, they will pass and ruin the system.' We can't think of a better endorsement for this book. And we believe, of course, that it is indeed possible to learn to swim—and to survive the course. In fact, we think that mastering certain skills along the way is just as important as getting across the lake to claim the prize—your PhD—on the other side.

The Problem: Saving an Old Master Painting from the Ravages of Time

To help illustrate some of the principles and suggestions outlined in this book, we'll be following a team of graduate students as they work together on an important project: saving a priceless Old Master painting from further deterioration. The robe of the Virgin Mary in the middle panel of *The Coronation of the Virgin* by Lorenzo Monaco (ca. 1414) is currently white. Technical examination has shown, however, that the robe was originally a deep pinkish mauve. A restorer can retouch the painting with red paint, but if the robe is still fading, a colour difference will occur; thus, elucidating the correct composition of the original paint, plus understanding the exact nature of the fading process, will be critical for carrying out a proper restoration.

Isabel, a chemistry PhD student, will be analysing the chemical composition of the paint. Her challenge will be to apply the analytical techniques currently available to study a sample from the painting, typically a tiny speck that is barely visible to the human eye.

Yousef is a PhD student in physics who will be focusing on calculations for the rate of fading in certain paint composites, as well as the important issue of whether it will be possible to reconstruct the original colours in the painting. Another aspect of Yousef's project will be to develop new analytical techniques for obtaining as much information as possible from the precious samples of paint.

Peter is working on his PhD in art history. His project will include the interpretation of the painting based on its use of colour, particularly when the colour is thought to have a religious or symbolic significance. The choice of colour may also be characteristic for this particular artist. The Virgin's robe, now white, though originally a deep pinkish mauve, for example, could be symbolic of her purity as the mother of Christ, whereas a purple hue would be a symbol of her royal nature as the queen of heaven. In order to solve the problem of the painting's continuing degradation, the team will have to work together and rely on each other's data. Communication, planning, and cooperation will be key to their success. All three are graduate students at the same university, albeit in separate research departments. Isabel has joined a well-established group run by a senior professor. Yousef works for a world-renowned professor in a large group with many PhD students and several Post-docs. Peter works as one of two PhD students for a young assistant professor.

To complete the project successfully, the team will need to draw on many skills they hadn't counted on using, including good communication, proper planning, and effective time management.

Zurich, Switzerland   Patricia Gosling
Amsterdam, The Netherlands   Bart Noordam

Authors' Note

More than ten years have passed since the 2011 publication of the second edition of *Mastering Your PhD: Survival and Success in the Doctoral Years and Beyond*. In the meantime, a great deal has changed in the ways we work and communicate with each other. Normally, technical innovations proceed in small steps. Extraordinary events, however, have a way of accelerating developments in all areas of life. The global COVID-19 pandemic led to rapid advances in biomedicine and communications. Effective vaccines were developed and tested in record time, and remote learning and online communication tools sprang up like dandelions. Although the concept of online learning has been around for two decades or so, overall acceptance of remote teaching, until recently, was low. Traditionally, learning took place in a classroom, and scientists from around the globe gathered at conference venues to share their latest research—despite the substantial investment of time and money involved. When the pandemic brought a halt to most face-to-face interactions, classroom teaching and onsite meetings were abruptly curtailed, and nearly everyone was obliged—seemingly overnight in some cases—to adopt remote communication tools. School teachers had to adapt their lesson plans for online classes, in addition to coping with the pedagogical and social limitations of remote learning. There was simply no other choice. Some adapted easily to the new tools for remote communication and collaboration, while others struggled with the unfamiliar technology. For the introverts and social-phobics among us, the ability to work or attend classes from home often felt like a blessing in disguise, whereas others struggled to cope with the forced solitude under lockdown conditions, including the lack of in-person meetings and the loss of social connectedness. In this respect, the pandemic acted as a magnifying glass, bringing into focus the many ways of being human, and the different paths each of us took to navigate a global health crisis that—at the time of writing—is still ongoing. As our rapidly changing world and the tools we use to communicate continue to evolve, we were inspired to create a new edition of this book, so we could include tips and advice on using the new learning tools and modes of communication during your PhD journey. To help illustrate some of these ideas, we've added a few more episodes to the 'Saving an Old Master Painting' storyline to show how Isabel, Yousef, and Peter successfully adapted the suggestions on remote learning and interactions in real-life practice. Even in a non-ideal world—which we have all been struggling through—these new episodes, along with additional tips and advice, illustrate creative ways for adapting our typical behaviours to fit a range of ever-changing circumstances.

September 2022   Patricia Gosling
                                                              Bart Noordam

Contents

1

Choosing a Research Group: Pluses and Pitfalls

Nothing great was ever achieved without enthusiasm.
—Ralph Waldo Emerson

Even before you get started on your PhD research, you will have already made a decision that will have a major impact on the success of your project, and perhaps even on your future career: you have chosen to work in a particular research group, and under the guidance of a specific thesis advisor or supervisor. While making this choice, you most likely spent a great deal of time thinking about your research project. You may have addressed questions such as: do I want to continue the type of research I did for my senior thesis/Master's degree, or do I want to explore a new field? Do I prefer doing experiments in the lab or do I feel more comfortable with a theoretical approach? No doubt you've had to think long and hard about these personal preferences.

However, there is another success factor for a productive and enjoyable research experience that you might not have considered: the group you'll be working in. Think of your team as occupying an island together in the most basic social unit of scientific research—science's nuclear family, so to speak—the research group. For the next few years, you will be 'stranded' on an island with this particular group of people—like it or not.

Every research group has its own unique chemistry and its own group dynamics. There are, nonetheless, some patterns to be aware of. It can be helpful to consider which type of group you're likely to be most comfortable

© The Author(s), under exclusive license to Springer Nature Switzerland AG 2022
P. Gosling and B. Noordam, *Mastering Your PhD*,
https://doi.org/10.1007/978-3-031-11417-5_1

in. Sure, you need to choose a group that's doing the science you want to do, but other factors like the size and nature of the group are also important. So, take some time to reflect on the others occupying your little research island, and what each of them is likely to contribute to the group effort, as well as your own research.

In this chapter we describe five archetypal research groups, ranging from a small group with a starting (assistant) professor as a supervisor, to a large group led by a senior (full) professor. We discuss typical advantages and disadvantages of the different types of groups, so don't forget to include these considerations in your decision process before you make a final choice. If you have already made your decision, you'll still be ahead of the game by being aware of some of the advantages and pitfalls of the type of group you'll be joining.

The Start-Up Group

Let's say you've joined a new group headed by a young assistant professor.[1] In this scenario you'll belong to the first generation of PhD students, and your advisor will likely be full of energy and eager for data—that you will have to acquire. *Caveat emptor*: young thesis advisors have the tendency to design overly ambitious research programmes. Plans may have to be simplified when reality sets in. In such a small group, you will have frequent and intensive interactions with your advisor, particularly because his or her career will depend on the success of the first generation of graduate students—that means you. An assistant professor's lack of experience in supervising students is usually compensated by the enormous amounts of time they typically spend with their small group. Moreover, your advisor will often have fewer teaching and administrative duties. Thus, more time can be devoted to working in the lab. Full of exciting new ideas, a starting professor often operates more like a senior partner than a boss. On the downside, there's the pressure to get tenured, and management and interpersonal skills don't come naturally to everyone.

---

[1] In this chapter, we use terminology from the US academic world to describe academic ranks: assistant, associate, and full professor. Each country has its own academic system with its own nomenclature. However, the career paths are similar. After completing a post-doctoral fellowship, one typically starts with a small (sub) group, as an assistant professor (in the US). In about five years or so, the group will have grown in size, and the supervisor is promoted to associate professor. When the group matures and reaches the status of a completely independent academic group, the supervisor is usually (but not always) granted academic tenure and promoted to full professor.

In such a setting it's crucial that you get along well with your advisor, so it's a wise idea to invest in the relationship. If there's no common ground and enthusiasm for your project at the outset, your life in the lab is bound to be rocky.

In a start-up group there will probably be just one or two other PhD students or Post-docs. The success of your projects will naturally be intertwined. The equipment and apparatus you'll need for acquiring data might even have to be constructed or set up. Lacking the infrastructure of an existing group, you will probably spend a lot of energy in building equipment, designing new models, or writing new computer code. It's a good idea to agree on a fair arrangement with your colleagues about how to share the output once it's time to harvest the data. Agreeing upfront on the order of authors' names on papers, for example, will prevent conflicts once the results start rolling in. Although this issue is relevant in any group, it surfaces often in start-up groups. These groups still lack clear publishing policies, while the output to be shared can be limited for the first generation of students.

How can you tell if the head of a start-up group has his or her act together, and if you're likely to make a success of your time there? After all, your young professor has no track record in guiding a group. Although success is never guaranteed, we offer two suggestions that might help you discover if a particular group is a good fit. First of all, consider your job application as a selection process that works both ways: the professor is looking for the best possible student, and you're searching for a group that suits you. Second, try to find out how the professor functioned as a Post-doc in a previous job. Was he or she already responsible for setting up new experiments and acting as a professor-to-be, or were they still working like a senior PhD student, acquiring data independently and executing suggestions effectively, but not doing more than that?

The 'Up-and-Running' Group

Around the time an assistant professor has delivered the first generation of PhD students into the world, he or she is usually promoted to the rank of associate professor. The new associate professor's initial research will have made some impact on the scientific community, and as a result, grant money is easier to come by, and the group is able to expand. The investments made by the first generation of PhD students are starting to pay off, so it might seem much easier for the second generation of students to do good research because the environment is much more conducive. Usually, however, the more established thesis advisor spends less time on research, since he or she is invited to give more lectures and to attend more conferences. Also, invitations to all kinds of committees are often eagerly accepted. How well the associate professor copes with this transition will depend on their organisational skills.

As a graduate student, you'll be working mostly with your fellow researchers. Guidance from your thesis advisor might be less frequent (as they may have less to do with day-to-day activities in the lab). Accept this reduced interaction with your advisor as a fact of life; after all, the ability to work well independently is a key career skill. Discuss how the two of you can have effective interactions when you *do* get together. You may have to make an appointment with your advisor to discuss progress on your research. This may be tricky, as your thesis advisor may not yet be used to scheduling discussions that were previously spontaneous. In a growing group it's as important to get along with your fellow graduate students as it is with your advisor. Those nearing the end of their studies can be a great help in kick-starting your research project (provided they have time and are willing). Be honest with them about your needs. In exchange for their help, you might offer to help them wrap up any remaining experiments for their thesis.

The 'Small-but-Established' Group

Having been a successful associate professor for several years, an academic scientist will usually be promoted to the position of full professor. With the rank of full professor and the hurdle of tenure cleared, their job is now secure. Some professors feel they're finally able to relax after many years of hard work. They may become more interested in the administrative aspects of running a research group, and their interest in academic research may start to fade. They've gained enough experience to keep a small group going, and periodically have decent or even great results to publish. When you're enrolled in such a steady but small group, you may have to work extra hard to generate enthusiasm for your project. Although such a group can get you a degree—a valuable asset in a future career in- and outside academia—it is not always the best place to start an academic career.

Interactions with your advisor may be infrequent, so proper planning of appointments with your advisor will be crucial under these circumstances. You'll not only have to plan these meetings but also prepare for them. Make a brief agenda of things you'd like to discuss and have your results ready in a presentable form. You most likely can expect more interaction and support from fellow graduate students and Post-docs, even if they're working on another area of research.

But not all small and established groups fit the description above. Plenty of full professors enjoy doing science so much that they remain deeply involved in the research at all times. They may choose to focus their attention on a small research group with just a couple of PhD students, and try to avoid administrative tasks as much as possible. These small enclaves of pure and intensive research can be wonderful and stimulating places for doing graduate student work. If you get along well with the professor and your fellow PhD students, you will likely thrive in this type of intensive environment, and your investment in group interactions will be much appreciated.

The Empire

Some successful professors allow their group to expand to immense proportions. Such groups can easily have 10–20 PhD students or more, along with several Post-docs. A good fraction of all PhD students conducts their research in such groups. Life for them is usually not bad at all, despite the fact that interactions with their 'famous' professor may be scarce to non-existent. Guidance comes from Post-docs in the group and more senior PhD students.

Not every topic investigated in those large groups will be a winner and some will fail altogether. But the availability of sophisticated instrumentation and a vast skill base should enable you to acquire data quickly. If your project fails, there are other projects you can fall back on. More than in any other group, interaction with your peers will determine your success and the amount of pleasure you get from your PhD research. Since there is a whole army of young researchers, you have the luxury of finding a few fellow students—and more senior scientists—with whom you click. You may even decide to work on a series of projects and share the results. Finally, be prepared for infrequent interaction with your famous professor. When she/he happens to drop by the lab, be prepared so you get the most out it.

Unfortunately, in some of these empires there may be a highly competitive and cut-throat atmosphere. During your introduction to the group, and certainly before you commit to working there as a PhD student, you should be alert to any lack of camaraderie and spirit of cooperation.

In some European countries, such as Germany, for example, full professors appear to run such empires. In reality, however, the professor often has a few lieutenants running smaller subgroups. When you work in such a smaller subgroup, the dynamics usually resemble those of a start-up group or an up-and-running group.

The Gardener

Though good scientists, once upon a time, the interest of these late-career researchers has waned. With their vast intellectual resources engaged in pursuits that most people only dabble in, they have over the years developed into an extraordinary gardener (or birdwatcher, or cook). They still maintain a research group—and even managed to get their grant renewed—but any good work that comes out of this lab is a result of the efforts of collaborators who haven't yet discovered that the 'gardener's' best years are behind them— or the occasional talented student or Post-doc who has the misfortunate to wander through, not having gotten word of their the lab head's scientific demise.

If you're not sure what you want to do with your life, or have a couple of years to kill, a stint in a lab like this can be just the thing. Just don't expect it to be the best possible start of a career in science.

Surviving in a Non-supportive Group

If you haven't yet chosen a research group, it's a good idea to give some thought to which kind of group you're more likely to function best in; how well you get on with your colleagues and advisor will have a big impact on your performance during—and after—your PhD project. You'll need to find a group that does the kind of science you're interested in, but also one that fits your personality.

Unfortunately, that's not a very easy thing for most of us to figure out. So much depends on the particular people in the lab—especially the person who runs it. You may think you're likely to thrive in a small group, only to find yourself very happy in a large one. It's easy to make a mistake. If you do—and even if you don't—there are likely to be times when you struggle with group dynamics, or when you feel you aren't getting the support you need. Here's how to get the most out of such a situation:

1. **Think positively**. Focus on the support that's available rather than sitting in isolation or frustrated about the support you're not getting or blaming

those who should be helping you. Use this time as an opportunity to develop some independence.

2. **Find the help you need**. No member of the group, including your advisor, can solve all your problems and fulfil all your needs. Some colleagues and advisors are better at designing new projects, others at debugging computer codes, still others at editing manuscripts. So, search around, and don't limit yourself to your research group; you might be, metaphorically, stranded on an island with these folks, but you do have a mobile phone and a fast internet connection. Don't share pre-publication data without your advisor's permission, but find the help you need where it exists, whether it's within or from outside your group.

3. **Identify your showstoppers**. It's not necessary to solve all your problems to make progress, at least not all at once. Set priorities and address the most important issues first. If you notice that you're spending a lot of time on problems that seem pressing but, after a careful analysis, aren't really all that important, focus only on those problems that really stop you from moving forward with your project.

If none of the above suggestions work, and you find yourself envying other students who seem to be working in a much more productive and pleasant group, it might cross your mind to change labs. This is a delicate issue that requires more time and space than we have to discuss in full detail. A few tips: double check how much greener the pastures are over there (is it really worth making a change?); identify the relevant procedures at your university (i.e., is changing labs common practice, or are you considering doing something that is unheard of?); finally, good diplomacy might improve the odds of making a successful move.

In summary, the type of group you work in can determine, to a large extent, the way you work, as well as the type and frequency of interaction you have with your PhD advisor. Your success will be influenced not only by having a good advisor (or a professor with a famous name), but also by your relationships with the other PhD students and Post-docs in the lab. So, try to evaluate the dynamics of the group you're in and identify your allies. Putting some effort into these relationships will help make your research projects more effective and your time in the lab more rewarding.

2

Getting Started

Don't judge each day by the harvest you reap, but by the seeds you plant.
—Robert Louis Stevenson

It's your first day in the lab. It's likely you're experiencing a range of emotions, from excitement to curiosity to anxiety. You'll be working in this lab and with this group of people, including your supervisor, for several years. This is Day One of a long commitment, so it's important to start off on the right foot. Perhaps you're so anxious to get going and prove your scientific mettle to others that you decide to do an experiment on your first day in the lab. But this would be unwise, to say the least. Give yourself a few days to get orientated, meet the people in your group, meet with your supervisor, and get to know the layout of the university, its facilities, and its graduate student services.

You will probably have been assigned a lab bench and a desk. In your early days in the lab, rather than jumping in with your first set of experiments, spend some time outfitting your workspace with the equipment and materials you'll need. Make your desk a comfortable and personal place to work—you'll be spending a lot of time there. You may also need to register for a university e-mail account and fill in forms for the departmental secretary, etc. Make sure you take care of all these administrative tasks before you get bogged down in your experiments in the lab. The following are some suggestions to make your first few days as a graduate student as smooth as possible and to help you get off to a good start. We discuss some of these issues in greater detail

P. Gosling and B. Noordam, *Mastering Your PhD*, https://doi.org/10.1007/978-3-031-11417-5_2

in subsequent chapters, but this brief sketch is meant to give you an idea of some important things to think about early on.

Become Familiar with Your Research Department

Use your first days as a graduate student to familiarise yourself with the inner workings of your department. If you haven't already done so, introduce yourself to the department chairperson, as well as administrative assistants, technicians, lab assistants, librarians, and other key personnel. This is not the time to be shy. Make a point of introducing yourself right at the start, so that people are not still wondering six months from now if you're a graduate student, an undergrad, a Post-doc, or a lab technician. Be courteous and open-minded when meeting people for the first time. The people with whom you'll be working will be important to you in more ways than you might realise, and first impressions count. You won't be able to work effectively unless you've familiarised yourself with your surroundings and met the key people around you, so be sure not to lock yourself up in the lab all day. Open up, mingle with your colleagues and make the effort to understand the ins and outs of your department. If you need information about the department or the university, ask senior graduate students and staff members. Be sure to introduce yourself to people who will be vital to your research, such as those responsible for ordering basic materials and equipment, operating technical equipment, and maintaining lab safety. In addition, be sure to familiarise yourself with lab safety and evacuation procedures. Know where to go for help when you need it. Perhaps most importantly, try to identify someone who might make a good mentor: a senior graduate student in your group, or a similar group, is an excellent choice. (see Chap. 8 for more information on how to find a mentor). This individual can help show you the ropes and provide valuable professional guidance throughout your tenure as a PhD student.

Formulate a Working Plan and Set up a Schedule

Before you even start that first experiment, you should establish a working plan and set up a reasonable schedule for yourself in which to complete the tasks in your working plan. It is best to do this with the help of your supervisor to be sure that you're working on the same goals. Divide your project into manageable phases and have a timeline for each phase. Be sure to set scheduled time off for yourself, because it isn't healthy, or wise, to work all the time. Once you've established your goals in conjunction with your supervisor and sketched out a timeline, keeping to your schedule is important so that your time is well spent (see Chap. 3 for a detailed discussion of goal setting and time management).

Maintain a Proper Lab Notebook

This may seem obvious, but it can't be stated often enough: a major factor in your research success will lie in your ability to keep detailed records of your experiments. Don't fall into the trap of thinking that only a neat lab notebook is a good lab notebook. Tidy tables of data are not enough. You must write everything down, including everything that worked and—just as important—all the things that didn't work. Don't be afraid to jot down random musings or thoughts in the margins. Forget about being neat. Meticulousness and completeness are more important attributes of a good lab notebook than perfect handwriting and tables drawn with a ruler. Make this a daily habit. Avoid rushing into an experiment without first writing down all the parameters—you won't remember every detail when the experiment is over. If you keep a proper record right from the beginning, making sense of your experimental data, and the logic behind your experimental design, will be a lot easier.

Establish Good Reference Keeping Practices

As you carry out your research, you will need to keep a working bibliography: a list of publications you will use as references for your research project. Start compiling this bibliography from day one and build it up as your work progresses. This will take away a huge part of the workload when you finally reach the stage of writing up reports, research papers, and your thesis. Once you have sufficient data for a preliminary report, get into the habit of writing up your work and submitting it to workshops or conferences. Presenting a paper at a conference or departmental meeting is good practice at this stage of your training. Establishing a track record through these types of presentations will help your career. Of course, every scientist's goal is to be published in high-impact journals. But that's something you can worry about further down the road. For now, the important thing is to stay focused and to keep good records of your work—both in the lab and in the library.

Dealing with Initial Ups and Downs

Life is full of ups and downs, and this is no less true than in your life as a graduate student. Experienced scientists know that research can be frustrating at times and not always go according to your well-made plans. Inexperienced scientists have a harder time managing their expectations and frustrations. So, expect to go through periods of stress and anxiety, whether due to work, study, or personal matters. Taking a little time off to relieve stress when the pressure gets too high is always a good idea. Don't feel guilty about having to take a break from time to time—you'll come back refreshed and ready to get on with your work. Most likely you've moved away from familiar surroundings to attend graduate school, leaving friends and perhaps family behind. Take the time to build a social network and your own informal support groups. These could consist of people from your research team or a group of peers, older colleagues or anybody you get along with. You'll need these people to share problems with and to go to for moral support when you need it most. Whatever you do, don't make the mistake of keeping problems to yourself. Everyone hits a difficult patch at some point in their graduate student years, so encountering problems is nothing to be ashamed of. Unresolved problems will not go away on their own. If you don't resolve them, they will keep you awake at night until you are so ground down by stress and anxiety that it will be hard to find your way back on the right path. Find someone understanding to talk to when things get too much to handle on your own. Lastly, while graduate student life can be demanding, take time out to have fun. There's more to life than the inside of a lab.

Connect with Others in Your Group or Department

- Use Wiki software (there are several options to choose from) to set up an online message board and collaborative space for your group or department, if one doesn't already exist, that contains relevant details and other practical information, such as how and where to order supplies, apply for travel funds, a list of upcoming conferences and meetings, and places to sign up for outings or other group activities. Wikis are a great way to build team spirit, work together more effectively, and share knowledge with the group.

- Start a virtual journal club. Spread the word, gather your group of participants, choose a line-up of speakers, and have each speaker present a journal article of their choice. Discuss the article in your online chatroom with colleagues from down the hall—or in another country.

Take a look at Chap. 15 for additional ideas on remote collaboration with your colleagues.

3

Setting Goals and Objectives

Give me a lever long enough, and a fulcrum on which to place it, and I shall move the world.
—Archimedes

As you work your way through the process of settling in, take some time (a few days if necessary) to write down your short- and long-term goals and objectives. Your ultimate objective is to write your thesis and obtain your doctorate, but that goal is years away, so you'll need to break things down into smaller steps. By dividing the different stages of your doctoral studies into manageable steps and committing them to paper, you'll not only avoid getting overwhelmed by the tasks ahead you will have a set of measurable and realistic goals towards which to work. One of the best ways to identify your goals is to start by creating an action plan. This type of activity typically involves the following steps:

1. **Clarify your goals and objectives.** First, look at the big picture and then break things down into shorter time segments. What do you want to have accomplished by the end of the first six months of graduate study? The first year? Sketch these goals out broadly, as they are likely to change over time. Now, write down your objectives for the next three months, and then fine-tune these for the upcoming month. Now that you've identified your goals, ask yourself two things: (1) are my goals measurable?; and (2) how will I know when I've achieved my goal(s)?

© The Author(s), under exclusive license to Springer Nature
Switzerland AG 2022
P. Gosling and B. Noordam, *Mastering Your PhD*,
https://doi.org/10.1007/978-3-031-11417-5_3

2. **Create a list of actions**. Now it's time to think about what you need to do to achieve your goals. Which limitations and constraints are you faced with in terms of time, know-how, equipment, material, etc.? Make note of any actions you can think of that will help you achieve your goals.

3. **Prioritise your action items**. Take a good look at your list. Prioritise the actionable points so that you first accomplish what is most efficient and the action that will most likely assist you in achieving your goals.

4. **Organise your actions into a plan**. Actions set into a time framework make up a plan. Make sure your plan is workable. Can you do the actions you have set up for yourself in the time frame you've allotted? Make sure you've ordered your actions into a logical sequence.

5. **Monitor and measure your progress**. On a regular basis you will have to monitor your plan and adjust if necessary. It's important to remain flexible and re-state your goals from time to time, as needed, and as you gain more experience with your project (for more details on how to do this, see Chap. 6: Monthly Progress Monitor). In the business world, some people prefer to follow the SMART method for establishing—and achieving—their goals:

- **S**pecific
- **M**easurable
- **A**ttainable
- **R**ealistic
- **T**ime-related

In other words, there is no point in setting a goal that you can't measure, can't attain, or isn't realistic. If, for example you're not physically fit, the goal of climbing Mount Kilimanjaro next week is specific and measurable, but unlikely to be either attainable or realistic in the time frame you've allowed yourself.

Effective Time Management

Once you've identified your short-term and long-term objectives, managing your time effectively will be key for sticking to your plan and attaining the goals you've identified. Most of us are familiar with that desperate feeling that time is slipping through our fingers, or that we don't have enough hours in the day to do all the things we need to do. Often that feeling of a lack of time has more to do with poor time management skills than with an actual lack of time. We all have the same 24 hours in every day. How we make use of them differs widely among individuals, and good time management is a major factor in successfully completing the goals you have set for yourself.

One useful tool in effective time management is to keep a record of your activities. You will be keeping a lab notebook of your experiments, of course, but it's also helpful to keep a written record on a daily or weekly basis of *all* your activities. This will help you analyse how you spend your time. The first time you start writing down all the things you do in a day, you may be shocked to discover how much of your precious time is being wasted.

You may also be unaware that your energy levels vary throughout the day and night. In fact, most people function at different levels of effectiveness during the 24-hour cycle. You probably know whether you're a 'morning person' or a 'night owl,' but do you know at which times of the day or evening your energy dips or peaks? Your productivity may vary depending on the amount of glucose in your blood, the length of time since you last took a break, routine distractions, stress, discomfort, or a range of other factors. Identifying your peak energy periods (and there are plenty of apps available for this very purpose) will help you use this time more wisely and focus on the things that count. By identifying your energy dips, you'll know when it's time to switch tasks, eat something to give you energy, or take a break and to get some fresh air.

Record Your Daily Activities

Keeping a record of your activities for several days will give you a better understanding of how you spend your time and when you perform at your best. Without modifying your normal routine or behaviour, take note of all the things you do (as you do them) over an entire day. Record your daily activities like this every day for a week. Every time you switch activities, whether it's reading e-mail, working in the lab, making coffee, sleeping, eating lunch, reading an article online, or attending meetings, note down when you do this and how you feel.

Learn from the Record

Once you have noted the way you use your time every day for a week, go back and analyse what you've recorded. It's not unusual to discover that you're spending a huge amount of time doing activities that are low on your list of priorities! (See the 80/20 rule below). You may also discover that you have more energy during some parts of the day and feel a bit listless and tired during others. Much of this variation in energy level depends on the breaks you take, the time and amount you eat, and the quality of your nutritional intake. Your written record will give you a basis for experimenting with these variables. Have you discovered that you have lots of energy in the morning and feel tired in mid-afternoon? Then get into the lab early and do your important thinking and/or experiments at that time. Use your low-energy time in the afternoon for more routine work, such as searching the literature or writing up your lab notes. An even better solution to beat periods of low energy is to get out of the lab and go for a brisk walk in the fresh air.

Another useful tool for getting everything done is to draw up a to-do list. This can be done daily or weekly, whatever works best for you. A to-do list is simply a list of all the tasks you need to carry out to reach the goal you have set for yourself. Once you've written your list, you can prioritise these tasks in order of importance.

There are people who make lists and people who don't. Perhaps you've never thought of yourself as a 'list' person before, but to-do lists are essential when you need to carry out several different tasks or different types of task, or when you've made several commitments that need attending to at the same time (multi-tasking). Don't make the mistake of believing you can juggle all this information in your head. If you find that you're caught out time and again because you've forgotten to do something, then you definitely need to keep a to-do list.

While to-do lists are a very simple tool, they are also extremely powerful, both as a method of organising yourself and as a way of reducing stress. Problems can often seem overwhelming, especially if they're left to rattle around in your head; or you may feel you have a huge number of demands on your time. This kind of stress may leave you feeling out of control and overburdened with too much to do. Creating a list (and crossing off/deleting things you've accomplished) can help relieve these feelings.

Prepare a To-Do List

The solution to feeling overwhelmed is simple: Write down the tasks you need to do, and if they're large, break them down into their component elements. If these still seem like too much to handle, break them down again. Go through this process until you've listed everything that you need to do. Once you've done this, run through your list and allocate priorities: 'A' (very important) to 'F' (unimportant). If too many tasks have a high priority, run through the list again and demote the less important ones. Once you've done this, rewrite the list in order of priority. When you're finished you will have a precise plan that you can use to eliminate the problems or tasks you are facing, one step at a time. Now you can tackle all these 'to-do' items in order of importance. This process will assist you in separating the important tasks from the many time-consuming trivial ones and keep you from 'tearing your hair out' in frustration from the growing mountain of things you need to get done.

Multi-tasking: Is It for You?

Multi-tasking is defined as doing more than one thing at a time, such as talking on the phone while checking your e-mail or eating lunch and reading the newspaper while recording data from an experiment. The more tasks we juggle to make the most of our time, however, the less efficient we become at performing any one task. The more time it takes to return to an interrupted task, the harder it is to remember where it was that you left off. Studies have shown that multi-tasking can greatly increase your levels of stress, so you'll have to decide whether it's the right approach for you. Some people are natural multi-taskers, others prefer to do one thing at time. Many people feel that multi-tasking, while a good idea in theory, diminishes their productivity and makes them work harder (but less efficiently) to keep up with all the things they're supposed to do. Advances in technology have made it even harder than ever to avoid multi-tasking, so you might try slowing down a bit and working on one task at a time to see how this affects your work—and your mood. Your concentration and productivity will likely increase, and it may be a relief when you no longer feel like you're running in a million directions at once.

The 80/20 Rule

Attributed to the Italian economist Vilfredo Pareto, the original concept of the 80/20 rule states that the relationship between input and output is rarely, if ever, proportional. When applied to your work, it means that only 20% of your efforts produce 80% of the results. Learning to identify the 20% that produces most of your results is the key to making effective use of your time. While simplistic in its conception, putting the 80/20 rule into practice is somewhat more difficult. So, how do you recognise the crucial 20%?

1. **Take a look at the people around you**. Twenty percent of your colleagues probably give you 80% of the support you need. They are your true advocates. Take time to learn from their example and to cultivate supportive relationships with them.

2. **Take a close look at your work**. Think about the scope of your project and ask yourself this: which 20% of my work should I be focusing on?

Are You Focusing on the 80% or the 20%?

Let's look at the above two statements in a bit more detail. The following are some indications of whether you're spending your time as you should. You're focusing on the 80% if the following statements are true:

- You're working on tasks *other people* want you to do, but you yourself have little or no stake in them.
- You're frequently working on tasks considered 'urgent'.
- You're spending time on tasks you're not particularly good at.
- Completing certain activities is taking much more time than you expected.

- You find yourself complaining all the time about how little you seem to be accomplishing compared to the effort you put in.
  You're focusing on the effective 20%, however, if:
- You're engaged in activities that advance your overall goals in the lab.
- You're working on tasks that you may not particularly like, but you understand that they relate to the bigger picture.
- You're asking for help with tasks you're not good at doing on your own.
- You feel a sense of accomplishment.

Implementing the 80/20 Rule

All of this might sound hopelessly simplistic, so if you're particularly sceptical, try applying the 80/20 principle for a few days just to see what happens. An increased awareness of the ways you work and the time you spend on a variety of activities will help you understand how to make use of this remarkably effective principle. You'll know it's working when you realise you have more time, are able to focus on what's essential, and can reduce the amount of time you spend on meaningless tasks or those that won't help you reach your goals.

Saving an Old Master Painting: Yousef Establishes a Set of Goals

Not one to waste any time, Yousef decides to start off on the right foot by establishing some goals and objectives during his first week in the lab. Some of these goals are not research related, such as familiarising himself with the department and the library and setting up his workspace. Even though he's anxious to conduct his first experiment in the lab, he takes time to write down some goals for his research. First, he'll need to do some background reading, as he knows very little about the chemistry and physical properties of paint pigments. Because he's a physics student, he'll also need to read up on art history, so he can put the project into context and have the vocabulary to talk to Peter, the art history PhD student he's collaborating with. In his notebook, or on his phone, Yousef sketches out his goals for the first month, as well as the three-month, and six-month mark, followed by a realistic action plan in the given time frame. Since he'll be using a relatively new technique for studying paint samples (secondary ion mass spectrometry, or SIMS), one of Yousef's goals is to conduct a thorough literature search on this technique. He also maps out an initial set of experiments and highlights possible pitfalls. Yousef is pleased he now has a plan to work with and goes out for a coffee. In the hallway, he runs into his supervisor and realises he hasn't discussed his plans with him at all. So, he tells his supervisor about his ideas and asks for a brief meeting to be sure his plans are in line with his supervisor's own ideas and vision. After some minor modifications they agree on the plan, and Yousef communicates to his teammates the things he plans to work on in the coming months.

4

How to Think Like a Scientist

Science, for me, gives a partial explanation for life. In so far as it goes, it is based on fact, experience, and experiment.
—*Rosalind Franklin*

By the time you've made it to graduate school, you should be well acquainted with the principles of the scientific method. Most likely the concepts have been drilled into you ever since high school biology class. Even so, we thought it would be a good idea to review some of the principles here, as they will form the core of your work in the lab.

Over the years, well-meaning friends and family members have probably asked this deceptively simple question: 'So what does a scientist do anyway?' or 'Tell me about your research.' You may or may not have a ready answer, depending on who is doing the asking and how much explaining you care to do. But imagine you're sitting around the dinner table and have been asked this question by a family member or friend, someone who knows nothing about scientists or the scientific method. How would you respond in a way that was clear and made sense to the non-scientist?

Perhaps the simplest and most accurate answer you could formulate is that scientists observe and measure the world around them. They gather information or data based on their observations, and when they think they have enough to answer the questions they have asked, they try to make sense of what it all means. During this process, most scientists use a reductionist approach. Let's say one scientist is studying a complex chemical reaction, another is investigating the foraging behaviour of the ring-tailed lemur, and a

P. Gosling and B. Noordam, *Mastering Your PhD*,
https://doi.org/10.1007/978-3-031-11417-5_4

third is researching the ocean currents around Tierra del Fuego. To make sense of these very complicated phenomena, each of these scientists must break down a specific aspect of the problem into simple components. These components are usually described as follows:

- Observation
- Constructing a hypothesis
- Carrying out experiments to test the hypothesis
- Formulating a theory

These four steps, taken together, are what is commonly known as the scientific method. If carried out correctly, the goal of the scientific method is to construct an accurate representation of the physical world. You may already have learned about the scientific method at some point in your career as a student of science, and while it may all seem very theoretical at this point, it will be important to keep these steps in mind as you go about your own research.

Because scientists are human and may be unduly influenced by personal and cultural beliefs and assumptions that alter their perceptions and interpretations of the natural world, the scientific method, if rigorously followed, can be thought of as an attempt to minimise bias. That doesn't mean, however, that the scientific method is without pitfalls.

Common Errors in Using the Scientific Method

Not Proving the Hypothesis by Experiment

Perhaps the most fundamental error a scientist can make is to mistake the hypothesis for an explanation of a phenomenon without having performed any experimental tests to verify the hypothesis. Sometimes, what we think of as common sense, logic, or intuition tempts us into believing that no experimental proof is necessary to prove the hypothesis because the answer seems so obvious. Consider a classic mistake made by the philosopher Aristotle, who many people consider to be the father of the scientific method. He emphatically stated that women have fewer teeth than men (probably to support his argument that men were superior), but never actually tried to prove this fact; he just used this misconception as a way to prove what everybody in ancient Greece accepted at face value anyway (that men are superior to women!). Of course, we all know that adult men and women have the exact same number of teeth—so don't fall into the same (not to mention outrageously biased) trap as Aristotle. Perform properly designed experiments to prove your hypothesis, rather than rely on 'obvious' assumptions.

Discounting Data that Don't Support the Hypothesis

Another common mistake is to ignore data that do not support your hypothesis. In the ideal situation, the scientist is open to the possibility that the hypothesis is either correct or incorrect. If, for example, the researcher has a strong belief that the hypothesis is true or false, even before collecting any experimental data, there may be a psychological tendency to find something 'wrong' with any data that does not support the researcher's expectations. It's hard to get rid of all our biases at once. The important point to keep in mind is that you need to treat all data the same way, and not assign a label a priori that it's either 'good' or 'bad'.

Over- or Underestimating Systematic Errors

A third type of common mistake occurs when systematic errors are either over- or underestimated. For example, many discoveries were missed by researchers whose data pointed to a new phenomenon, but the data were mistakenly attributed to 'experimental noise'. Conversely, data that are part of the normal variation of the experimental process could be interpreted as evidence of a new discovery.

How can this tendency towards bias be reduced? An important check on bias is to promote open communication among the members of a scientific field in the form of publications and conferences. In this way, the biases of any one individual will most likely be cancelled out as other scientists try to reproduce their results. In time, a consensus may develop in the research community as to which experimental data has withstood the test of time.

Fact, Theory, Hypothesis—What's the Difference Anyway?

These terms are not interchangeable, even though they are often treated as such in popular usage. For a scientist, each of these terms has a specific definition:

A **fact** is something that is known to be true. Fire burns wood into ash. Under standard pressure, water boils at 100 degrees Celsius.

A **theory** is a conceptual framework that can be used to explain existing observations and predict new ones. For example, the path the sun follows as it crosses the sky can be explained by the theory of gravity.

A **hypothesis** is a working assumption. Usually, this assumption is formulated *before* experiments are carried out to test it. If the hypothesis holds up against existing and newly obtained data, the scientist may formulate it as a theory.

Is There Ever a Situation Where the Scientific Method Is *Not* Applicable?

A frequent criticism of the scientific method is that it cannot accommodate anything that has not already been proved. This argument points out that many things thought to be impossible in the past are now everyday realities (such as flight, for example: Two hundred years ago, it was believed impossible for humans to fly in the air). This criticism, however, is based on a misunderstanding of the scientific method. When a hypothesis passes the test, it is adopted as a theory, which can correctly explain a range of phenomena. This theory, however, can always be falsified by new experimental evidence. But it is not necessary for the hypothesis to have been previously proved for the scientific method to work.

Ockham's Razor

In the fourteenth century, William of Ockham proposed the principle of Ockham's Razor as follows: *Pluralitas non est ponenda sine neccesitate*. This can be translated as: Entities should not be multiplied unnecessarily. In other words, 'keep it simple'.

Suppose, for example, you have two theories that predict the same thing. In this instance, the principle of Ockham's Razor can come in handy. Here are two sample theories that describe the same phenomenon:

1. The tides on earth are influenced by the position of the moon.
2. The tides on earth are influenced by the position of the moon, which is determined by the force of powerful extra-terrestrial beings.

Both theories make identical predictions, but Ockham's Razor would discount the second theory as containing unnecessary information. The simpler theory works just as well. Ockham's Razor does *not* guarantee, however, that the simplest theory will be correct, it merely establishes priorities.

A Final Comment

Biases aside, the scientific method is the best approach we've come up with to accurately answer questions about the physical world we live in. Without the scientific method, we might still believe in the idea of spontaneous generation (i.e., that flies, for example, are 'born' from rotten meat), a theory disproved by Françesco Redi and Louis Pasteur in an ingenious set of experiments that applied the principles of what later came to be known as the scientific method. As a result of his experiments Pasteur concluded that there is no life force in air, and organisms do not arise by spontaneous generation (from rotten meat, for example). Pasteur recounted his famous experiment with the memorable phrase: 'Life is a germ, and a germ is Life. Never will the doctrine of spontaneous generation recover from the mortal blow of this simple experiment'.

Saving an Old Master painting: Isabel forms hypotheses about the whitish, transparent inclusions in the red paint

To prepare for her work in the lab, Isabel has been doing a lot of reading in the library and online about the chemistry of paintings and some of the problems that paintings undergo after several centuries of being exposed to light, air, humidity, and extremes of temperature. In her reading, she has discovered that some degradation can be the result of the formation of leadsoap aggregates of certain pigments, including red, lead-containing paints. These lead–soap aggregates can expand and remineralise, thereby changing their chemical composition. Because of some other evidence found on the painting, such as the breakup of the overlying paint layer and whitish opaque material protruding through the surface of the painting, Isabel hypothesises that aggregates have formed and remineralised to lead carbonate, a remineralisation product. If this is indeed the case, she further hypothesises that the red lead reacted with fatty acids released during the ageing process of the oil-binding medium. To prove her hypotheses, she decides to analyse the paint sample with a number of imaging techniques, including FTIR, SEM/EDX, and SIMS, selecting the best technique for determining whether lead-soap aggregates have formed, if they have subsequently remineralised, and if the remineralisation product is in indeed lead carbonate.

5

Designing Good Experiments

Progress in science comes when experiments contradict theory.
—Richard Feynman

In the previous chapter we talked about what it means to 'think like a scientist' and how to successfully apply the scientific method to your work. A critical feature of this process involves testing your hypothesis with experiments. Good experimental design for each and every experiment you conduct will greatly enhance your chances of success in the lab. Even if you obtain a negative result, a well-designed experiment will increase your confidence in your work and the reliability of your data. Designing a suitable experiment to test a hypothesis takes ingenuity and skill. Whether or not your experiment requires sophisticated equipment, a number of features are common to all well-designed experiments:

1. **Discriminating between different hypotheses**. In a well-designed experiment you should be able to discriminate between two hypotheses. In a poorly designed experiment, you may obtain results that support more than one hypothesis. If you carry out an experiment and discover that this is the case, then it's time to start over.

2. **Replicating your results**. When you carry out your experiment several times, are you able to replicate your results? If not, there is a serious flaw in your experimental design.

© The Author(s), under exclusive license to Springer Nature Switzerland AG 2022
P. Gosling and B. Noordam, *Mastering Your PhD*,
https://doi.org/10.1007/978-3-031-11417-5_5

3. **Controlling the variables.** Experiments must be well controlled against each of the variables tested. It's important to eliminate the possibility that other factors in the overall experimental setup are producing the effect you observe, rather than the factor you're interested in studying.

4. **Methods of measurement.** Your methods of measurement must be reproducible from day-to-day, between different researchers in the same laboratory, and between different laboratories. Many scientists do not consider a result to be valid unless another researcher has reproduced the same results. Maintaining the accuracy of your instrumentation and exercising quality control in your laboratory practices are critical.

5. **Blinding.** In Chap. 4, we talked about experimenter bias. It's possible for researchers to unconsciously 'fiddle' their data to get the result they hoped to obtain. In order to avoid this type of unconscious bias, it may be desirable to carry out experiments in which you don't know, for example, which compound is being tested in which laboratory rats, or which chemical reaction is taking place in a particular reaction vessel. This method is routinely used in clinical trials of drugs, in which both the doctor and the patient are unaware of the treatment they're receiving. These so-called 'double-blind' studies are meant to avoid bias on the part of doctors and the patients as to the efficacy of a drug. The placebo effect has been well established.

6. **Accuracy and precision.** In almost any type of experiment, you will most likely have to measure something (e.g., the rate of a chemical reaction, the glucose concentration in blood, the orientation of molecules in reactive scattering). Therefore, it is critical that you know both the accuracy and precision of your measuring device. Measuring the length of a fruit fly with a yardstick, for example, would not be very accurate, although it would probably be quite precise. These two terms are not synonymous, and it's important to understand the difference. *Accuracy* means the ability of a method to give a true measurement on average. *Precision* is a measure of the method's reproducibility. Your method of measurement should be both accurate and precise (i.e., have a low standard deviation). Sometimes, one of these factors is more important than the other. If you are measuring changes over time of a particular quantity, it's more important to have a precise method of measurement than an accurate one. Accuracy and precision are important factors in your experimental design, as they will

determine the reliability of your data, as well as the number of significant figures you can use in reporting your results.

Good Science and Good Experimental Design Go Hand in Hand

To assist you in designing good experiments, we suggest you follow these three steps:

1. **Define your objectives.** What is it that you're trying to test in this particular experiment (i.e., which question are you trying to answer)?

2. **Plan your strategy.** How will you achieve this objective? What is the size and scope of your experiment, and how many times will you try to repeat it?

3. **Pay attention to experimental details.** Sketch out the details of your experiment. Which tools and equipment do you need? How much time will your experiment take (one hour, one day, one month?)

If you're a biologist who's conducting experiments in a population of plants or animals, or a clinician doing a clinical trial with human subjects, a statistician would tell you to think about the statistical tests you will need to perform to analyse your data. Knowing this information will help you plan your experiment from the start. For example, you'll need to know beforehand how large your study population needs to be to give you enough statistical power for your analyses.

Once you have identified your objectives and formulated a hypothesis, you need to define the variables you'll use for testing your hypothesis. A well-designed experiment should have only one independent variable. If you change more than one variable at a time per experiment, you will not know which variable is causing the effect you observe. Keep in mind, however, that some variables are linked and may influence each other to create the same effect. Initially, try to choose variables that you think act independently of each other.

Make a list of answers to the questions you're asking. This can be a list of statements describing how or why you think the things you've observed act as they do. These questions should be stated in terms of the variables you've previously identified. Normally, you should formulate one hypothesis for each question you have. And you must do at least one experiment to test each hypothesis.

Design Experiments to Test Your Hypothesis

The next step is to design an experiment to test each hypothesis. In order to do that, you'll need to make a list of the things you need to do to answer each question. The list you create will be your experimental procedure. This procedure should include the appropriate methodologies, technologies, and equipment. For some types of experiment, you will also need a 'control' to act as a reference. A control is an additional experimental trial or run. It's a separate experiment, performed exactly like the others, except that no experimental variables are changed. A control is simply a neutral 'reference point' for comparison that allows you to observe the effect of changing a variable by comparing it to a situation in which you change nothing. In practice, dependable controls can be difficult to design and can be the most difficult part of your experimental setup. Without a proper control, however, you cannot be sure that changing a particular variable has caused the effect(s) you observed.

1. **Prepare your materials and equipment**. Make a list of all the things you will need to perform the experiment, including chemicals, glassware and reaction flasks, instrumentation, etc. Gather the materials and equipment you need and make sure everything is functioning properly.

2. **Record the data**. Experiments are often carried out in a series. For example, you can perform a series of experiments by changing one variable by a different amount each time. A series of experiments is made up of separate experimental 'runs.' During each run you record a measurement of how much the variable affected the particular system you're studying. For each separate run, you change the variable by a different amount. These changes will produce a different effect on the system. You measure this response and record the data in a table or chart. The data you collect at this stage is 'raw data' since it has not yet been analysed or interpreted. Only when raw data are analysed can your findings be described as 'results'.

3. **Record your observations**. Record your observations during the experiment, remembering to take note of any problems that crop up. Don't forget to keep careful notes of everything you do, and everything that happens from the beginning of the experiment to the end (no matter how trivial or unimportant it may seem at the time!). Careful data collection and observation are crucial to the scientific method. Your observations will be invaluable later on when it's time for you to draw conclusions, as well as for spotting any sources of experimental error.

4. **Analyse the raw data**. Now you need to perform the necessary calculations to transform your raw data into the numbers you'll need to draw your conclusions. For example, if you weighed an empty reaction vessel, this weight is recorded in your data table as 'wt. of round-bottom flask'. You then added some sample to the container and weighed it again. This would be entered as 'wt. of flask + sample.' To complete the calculation section, perform the calculation to find out how much sample was used in this experimental run: (wt. of flask + sample) − (wt. of flask) = wt. of sample used. This is obviously a very simple example, but you get the idea! Nothing is too trivial to record in your lab notebook. Never rely on your memory of procedures and calculations.

5. **Draw conclusions**. Based on the trends found in your experimental data and your observations, try to answer the questions you asked at the start of the experiment or set of experiments. Is your hypothesis correct? This is the time to assess the experiments you performed. Ideally, you should be able to evaluate the relationship between the predicted result contained in the hypothesis and the actual results and reach a conclusion as to whether the explanation on which the prediction was based is supported—or not.

Other points to consider when summarising your conclusions:

- If your hypothesis is *not* correct, what could be the answer to your original question?
- Summarise any difficulties or problems you encountered while doing the experiment.
- Do you need to change the procedure and repeat your experiment? What would you do differently next time?
- Make a list of other things you learned.
- Try to answer any other related questions. Interpretation of the data may lead to the development of additional hypotheses, the formulation of generalisations, or explanations for natural phenomena observed.
- Record any experimental errors, including possible explanations.

Finally: Can You Trust Your Results?

You've designed your experiment properly (you hope) and carried it out according to the methods and procedures you devised. If you did not observe anything different compared to your control, the variable you changed may not have had any effect on the system you're investigating. If you did not observe a consistent, reproducible trend in your series of experiments, experimental errors may be affecting your results. The first thing to check is how you're making your measurements. Is the method of measurement questionable or unreliable? Perhaps you're reading a scale incorrectly, or the measuring instrument has not been properly calibrated. If you can determine that experimental errors are indeed influencing your results, carefully rethink your experimental design. Review each step of the procedure to locate sources of potential errors. If possible, have a more senior scientist or fellow graduate student review your procedures with you. As the designer of the experiment, you can sometimes miss the obvious!

Spotting Random Errors

If your measurement methods are not the cause of the error, try to determine if the error is either systematic or random. Random errors are easier to spot. They result in non-reproducible data that don't make sense. In this case, experimental runs with the same combination of variables, and even the control itself, cannot be duplicated. Some randomness is always present in nature. No two measurements are exactly the same, so you must decide if the differences in your data can be explained by natural variation.

A random error may be occurring because you're doing something differently in each trial. For example, if you have not thoroughly cleaned your glassware or instruments, some of the chemicals or DNA or peptides you used may be carried over from the last experiment. Various statistical tests might help you determine if the difference between experimental runs is due to randomness, or to your experimental procedures.

Identifying Systematic Errors

Systematic errors are more difficult to identify. At first glance, your data and results appear to be consistent and reproducible, so you're unaware that something is causing all your measurements to be off by the same amount. For example, if you were not aware that your balance was off by 3 mg, all your measurements of weight would be off by 3 mg. This type of systematic error will affect all your data by the same amount. One way to check for systematic errors is to run experiments that have a different design, but should give you the same answers. It's can be a good idea to carry out different types of experiments to cross check your results. Another way to locate errors is to have an independent investigator repeat your experiments.

Recognising Linked Variables

Your results may be invalid if your variables are not independent of one another, and you have failed to notice this relationship. Variables are only independent if they produce their effects separately from each other. In other words, if a variable is independent, changing it will not influence the effects produced by another variable.

What if Your Experiment Hasn't Worked Out as Planned?

No matter what happens, whether your experiment was a success—or not— you will have learned something. Because science is much more than just getting 'The Answer'. Even if your experiments don't answer the specific questions you asked, they will give you ideas that can be used to design additional experiments. Knowing that something didn't work as expected, is actually knowing quite a lot. Thus, unsuccessful experiments are an important part of the process in finding an answer to your question. Incorrect hypotheses have a value of their own, as they can help point the way towards further investigations.

If you're struggling, or having doubts about your experimental technique, you are not alone! For collaborative projects or helpful insights from your colleagues, consider sharing your (online) lab notebook with others in your group via popular project management platforms such as Google Docs, Trello, or Workflowy.

6

Charting Your Progress Month by Month

If you fail to plan, you plan to fail.
—Benjamin Franklin

Having dealt with the initial problems and uncertainties of the first few months of life as a graduate student, you're most likely feeling more comfortable in your new environment. Your computer is up-and-running, you've acquainted yourself with the working habits of your institute, and you know how to acquire the data you need (e.g., by doing experiments in the lab, conferring with your supervisor, or data mining via online searches or in the library). You've been working long hours and may even have sacrificed numerous weekends to get more done. Some time ago, in agreement with your supervisor, you sketched out the targets you plan to meet in the first year. They seemed reasonable on paper, and the planning looked realistic. You're looking forward to making your first scientific breakthrough. Nevertheless, the gap between the actual progress you've made and the targets you have set for yourself is growing wider every day. Somehow, despite all your hard work, you're not approaching your goals. Soon, this may even become your daily mantra: *Why am I not approaching my target when I'm running so fast?*

The problem with getting into this mindset, is that it is often difficult to recognise the patterns that slow or even inhibit your progress. Your first thought may be that you need to work harder to catch up, but very likely you have already discovered that this approach doesn't work. It may feel like

© The Author(s), under exclusive license to Springer Nature
Switzerland AG 2022
P. Gosling and B. Noordam, *Mastering Your PhD*,
https://doi.org/10.1007/978-3-031-11417-5_6

you're taking one step forward and two steps back. Perhaps it has already crossed your mind that it might be easier to just quit. Fantasies of starting a 'bed and breakfast' in the south of France begin to drift through your mind. As an undergraduate, when your targets typically had a time span of a few weeks, you were doing fine. Now you're starting to discover that there is more to reaching annual goals than adding up hundreds of daily steps. You need some sort of monthly evaluation to bridge the wide gap between your one-day and one-year plans. This chapter provides you with a tool to help: the Monthly Progress Monitor.

Monthly Progress Monitor: Four Questions to Keep Yourself Goal-Orientated

For a monthly evaluation scheme to be effective, we believe it should be simple and easy to use. So, we developed a form (see the end of the chapter) that asks you to answer only four questions. The scheme has been tested extensively in various research groups in several countries, including the Netherlands, Denmark, and the United States. The four questions are:

1. Of the results I obtained last month, which are the most important?
2. Did I deviate from last month's planning? If so, why?
3. What are my most important goals for the upcoming month?
4. What do I need to do to reach these goals? Which potential hurdles might I face, and how do I overcome them?

These questions are meant to help you understand the patterns inherent in your working style, so it's very important to fill in the answers to the questions, particularly prior to meeting with your supervisor. We suggest that you fill in the form every month throughout your entire doctoral studies. More frequently is not practical and too much of a burden on your time. Less frequently would make the targets too vague for direct action and practical solutions. Based on our experience we have found that answers to the four questions will reveal the following:

Question 1—At first, you may think you accomplished so many things last month that it will be hard to summarise them. But if you focus on the really important issues that pertain to your doctoral studies, you'll be able to come up with a short list. It may be a bit shocking to realise how much time you spent last month on issues that aren't on the list of major contributions towards your thesis (see the 80/20 rule, Chap. 3). Most newly minted PhDs agree, with the power of hindsight, that they could have obtained their degree much faster if they had only followed paths that were productive. Of course, the very essence of research is that you do not know the answers beforehand, nor the productive pathways, and not all your lines of inquiry will work out. But prioritisation of your work can do no harm. If you're lucky, you will pick more productive approaches than just randomly throwing darts at a dartboard (while blindfolded) and trying to hit your targets.

Question 2—Now for the tough part. Compare the answer to Question 1 of this month's evaluation (what you have done) to the answer to Question 3 of last month's evaluation (what you planned to do). Most likely you will have accomplished only a small fraction of last month's ambitious plans. Record your thoughts on why you weren't able to do more. While it may seem obvious, by re-reading the answers to Question 2 from the last few months you'll start to see patterns in how you work. Recognising the problems in your working style is very often the first (and most difficult) step in finding the solution.

Question 3—Naturally, as ambitious as you are, you have numerous plans for the coming months. But having pondered the answers to Question 2, you have perhaps become more realistic—and wiser. Your list of goals for next month will now be rather short, otherwise you won't be able to finish them and you'll end up with the same long list of projects you've started but not yet finished. In fact, what you're doing now is prioritising your projects for the next month. Because prioritisation seems so obvious, it's often neglected. The lack of proper prioritisation is one of the main pitfalls on the road to getting your PhD. Be sure to make your goals for the coming month truly action-able. For instance, an action such as *understanding more about the chemistry of paintings* is too vague. A more measurable target might be: read and under-stand three published articles (such as ...) and Chaps. 2, 3, and 5 of the book 'The Chemistry of Paintings'. After readings these articles and book chapters, try to formulate a hypothesis about what's going on in the painting you're investigating and a possible approach for testing that hypothesis. It may take a few extra minutes to come up with such a refined plan of action, and don't hesitate to consult your supervisor or a more senior scientist if you need guid-ance. But stick with it, as the effort will pay off. Perhaps your original plan led to a conversation with your supervisor that went something like this:

'I've been reading some literature about the chemistry of paint pigments.'
'So, what are you going to do next?'
'Actually, I have no idea. Maybe I should read more chapters in this book. I found the biochemistry sections to be rather difficult.'

'I don't believe the biochemistry is relevant at this point, maybe you should take some mass spectra of your samples and analyse those first.'

In short, you've used up quite a bit of your valuable time by just reading. Your supervisor is correcting this approach, and he or she is again taking the lead in your project. Instead, if your initial discussion a month earlier had been more comprehensive, you would have read the relevant chapters and come up with some suggestions for the next steps. Even if you suggest the wrong set of actions, you'll have learned something from thinking about the path forward. Taking a more pro-active role in your research starts with making an actionable to-do list for the month ahead.

Question 4—You're a pro when you master the answer to this question. Knowing the potential hurdles and obstacles in the projects you have selected to work on in the coming month is far from easy. Finding solutions to circumvent these hurdles is even more difficult. But spending time on fore-seeing the hurdles and taking proper measures to keep them from stopping you in your tracks is very rewarding. *Staying ahead of the problem* is a skill that will not only make your PhD a success, but one that will help you in all your future endeavours. In fact, avoiding potential pitfalls in future projects will save you an enormous amount of time. Make sure to properly use the time you saved by being more efficient. Do not run blindly onto another track. Balance this extra time between: (a) working fewer hours (an hour in the gym can be a more efficient use of your time than another hour in the lab); (b) thinking of other potential hurdles and how to circumvent them; and (c) doing a little extra work on the relevant problems.

What You Can Learn from Completing the Monthly Progress Monitor

Once you've been using the Monthly Progress Monitor for a few months, you should go over the old forms again. If you're using the paper version, file them properly in a dedicated folder—they make a great record of your research progress—or store the digital versions on OneNote, for example, and share access with your supervisor and/or others on your team. You will typically find that:

- At first, planned work in the month-to-come tends to exceed the progress made in the previous month.

- After using the Monthly Progress Monitor for a few months, your planning will become more realistic.
- Expectations and goals become aligned, resulting in fewer conflicts.
- The general progress of your project has improved because of the timely identification of possible hurdles.

Above all, try to be honest in evaluating your progress. An honest evaluation can help you identify patterns and obstacles that are slowing you down. If you have the courage to do so (as it will necessarily involve discussing weaknesses in your working style), discuss with your supervisor the 'big picture' that comes out of this exercise. Alternatively, you could consult your mentor, a friend, or a sympathetic colleague if you're feeling stuck, or need advice on how to move forward.

> Take a look at Chap. 8 for suggestions on finding a mentor, and why a mentor can be a valuable asset (alongside your supervisor) when it comes to guidance and advice.

Monthly Progress Monitor

Name of PhD student: Name of supervisor:
Date: Previous meeting:

1. Of the results I obtained last month, which are the most important?
2. Did I deviate from last month's planning? If so why?
3. What are the most important goals for the upcoming month?
4. What do I need to do to reach these goals? What are the potential hurdles and how do I overcome them?

Suggested date for next month's meeting:

General agreements:

a. *PhD student fills out form prior to meeting with supervisor*
b. *During the meeting, the answers to the four questions above are completed*
c. *Supervisor gets a copy of the completed form after the meeting*

7

Dealing with Setbacks

If you know you are on the right track, if you have this inner knowledge, then nobody can turn you off…no matter what they say.
—Barbara McClintock

By now, you have settled comfortably into your lab routine: you've established your goals and objectives, your research project is well underway, and you've been carrying out experiments for several months. You feel good about your progress and are convinced you're on the right track. You've mastered the concept of thinking like a scientist by working through the classic progression of hypothesis, experiment, and results, and you are feeling confident that you have a good handle on your project and your life as a graduate student. Each month you've faithfully completed the Monthly Progress Monitor, and you're keeping open the lines of communication with your supervisor and colleagues. So, everything is wonderful, right? Wrong. Because one day you wake up and realise that *nothing is working*.

Your carefully planned experiments are not giving you the results you've expected or need. Your cell cultures have become contaminated for the umpteenth time. The PCR machine or the HPLC or the UV spectrometer breaks down... again. You can't get your chemical compounds to crystallise, or you injected your laboratory mice with a mislabelled syringe, and they all die. Weeks or months of data are lost. To top it off, you find out that you've made a mistake in your statistical calculations and a year's worth of experiments are worthless. Gather a group of seasoned scientists together in one

P. Gosling and B. Noordam, *Mastering Your PhD*, https://doi.org/10.1007/978-3-031-11417-5_7

room and they will tell you horror stories like these and more. When something like this happens to you (and it likely will), what do you do, and how do you cope?

Setbacks in the lab—and in life—are inevitable. It's how we deal with them that will turn a setback into an opportunity for growth. If you think of setbacks as not being failures, mistakes, or wrong turns, but rather a chance to learn and grow, you will be much better equipped to get yourself back on the right track.

The Cold Reality of Trial and Error

Your experiment worked beautifully the first time you tried it, but now you can't repeat it, no matter what you do. You're tearing your hair out and losing sleep. The experiment was flawlessly designed, elegant in its construction, and the results you obtained fit beautifully with your hypothesis. And now nothing, nothing you do produces the same result.

Now might be a good time to remind yourself that much of science involves trial and error. It's an unavoidable fact that you will make a number of errors and false starts before you're able to reveal even a glimmer of truth. Science proceeds in fits and starts. There are no quick fixes, no overnight successes. The very nature of scientific discovery requires that it proceed in its own time. Experiments that are obliged to follow the timing of the natural world (bacterial growth, metabolism, chemical reactions, nuclear decay, etc.) cannot be hurried, even if you do have a deadline looming.

Keep in mind, too, that it's virtually impossible to avoid surprises, unexpected results, and setbacks in the lab. If all research proceeded without a hitch, scientists the world over would be able to skip the endless rounds of experiments and go straight from hypothesis to publication without breaking a sweat. So, take heart from the fact that many things will go right, and try not to focus on all that's gone wrong. Also, it may help to remember that a *negative* result from one—or even a series of experiments—may be just as valuable as a *positive* result.

Part of your training in learning how to 'think like a scientist' involves dealing with the inevitable setbacks that will occur and learning to cultivate the virtue of patience. And you must do this while developing a fine sense of balance: keeping the big picture in mind as you revel in the small steps forward and cope with setbacks along the way.

So, while the overarching goal of your lab may be to find a cure for cancer or to understand the underlying mechanisms of a genetic disease, this goal is not going to be reached overnight or even next year, or very likely not even five years from now. Scientific progress and so-called 'breakthroughs' typically occur only after decades of hard, painstaking work.

But notwithstanding all this, if you feel that setbacks in the lab are taking their toll on your confidence, a logical approach to identifying the problems will put setbacks in perspective, turn them into learning experiences, and boost your confidence.

Identifying the Setback(s)

A setback can be defined as an event or occurrence that prevents you from achieving your goal (a failed experiment, a string of failed experiments, contaminated cell cultures or animal models, badly calibrated machinery leading to loss of data, etc.). To help you identify what your setback is (and how to recover from it) try answering as best you can the following questions (it helps to write them down).

1. What is your setback?
2. What (in your own words) have you failed to achieve?
3. What mistakes have you made?
4. Who, if anyone, has disappointed you? (colleagues, your supervisor?)
5. What do you regret doing or not doing (in relation to the setback)?

As you review your answers, explore your negative emotions and thoughts (are you feeling overwhelmed, that you're in over your head, sad, frustrated, angry?). Try not to take it personally. The fact that you're experiencing a setback doesn't mean that you, as a person, are somehow a failure.

Take Action

When you're depressed or frustrated about your circumstances, it's easy to turn to self-destructive activities such as over-indulging in alcohol, junk food, or excessive TV watching. This is not to the time to eat your body weight in ice cream or drown your sorrows in drink.

First, ask for support from those around you. Talk to colleagues, friends, and family about your concerns. Then try to find concrete ways that will get you back on track in the lab. Talk to older scientists, your mentor, and your supervisor. Chances are, all of them will have similar stories of false starts, frustrations, and setbacks. Hopefully, the setbacks you're experiencing are just bumps in the road and not full-blown obstacles that will prevent you from reaching your goal. When you feel better, try to put things in perspective. Frustrations and setbacks are a part of life. It's how we deal with them that counts.

Tips for Recovering from a Setback

You're halfway to a solution once you acknowledge you're stuck, and that something must be done to get your research back on the road to recovery. Here we offer a few practical tips for getting past the point of acknowledging there's a problem and moving towards a solution.

1. **Take care of yourself**. Most likely you've been putting even longer hours into your work to recover from your setback. But running yourself down physically and exhausting yourself mentally will only increase your misery and feeling of hopelessness. Regular meals, plenty of sleep and lots of stress-reducing exercise will go a long way towards getting you back on track.

2. **Think outside the box**. Creative solutions require out-of-the-box thinking. So far, what you've been doing has not been working, so it's time to take another tack. You may need to change your experimental plan and work on something different. Perhaps you will even need to take more time to finish your PhD work than you had planned. Perhaps you'll need input from a Post-doc or more senior scientist to help you get your complex experiments going again.

3. **Involve others**. Although it will take some courage to admit to others that you're experiencing serious setbacks, you need the support of people around you more than ever. Friends and colleagues can offer the moral support you badly need. Having someone who can listen to your problems (and even better, offer useful advice) will be invaluable on the road to recovery.

4. **Cultivate the art of patience**. It takes time to recover from a setback. Your problems will not disappear overnight, so it's important to cultivate the art of patience and being kind to yourself. Accept that it will take time to get back on track again. Focus on the things that are already going better,

rather than on all the things that still need to be done. Take one step at a time and remind yourself that 'Rome wasn't built in a day.'

A Final Thought: Should You Stop Altogether?

You wake in the middle of the night—night after night—and a single thought weighs heavily on your mind: 'Things are not working out. Grad school is not what I thought it would be. I don't really want to be a scientist after all. Maybe I should quit altogether.'

Life is fluid. Peoples' ideas about themselves and their place in the world change from one year to the next. You started graduate school all fired up, anxious to get on with your research and your career, but now you're not so sure anymore. It's not just a question of a few lingering doubts that drift through your head at the end of a bad day. We all have those. But if they are persistent and impossible to ignore, you may need to consider the possibility that going all the way to the end of your PhD may not be right for you. Should you quit? The answer to this question will be different for everyone. So much depends on who you are as a person, your goals in life, and how much you've already invested in the process of getting a PhD. If you're still in your first year and you're having serious and persistent doubts that you've made the right choice, it would be a good idea to talk things over with someone you trust (not your supervisor at this point—why set off alarm bells if you don't have to?).

It's never too late to change your mind and move on to something else. If you're more than halfway toward your PhD, however, your decision will be more complex. You'll have to weigh the balance of losing the time you've already invested, with your desire to change careers. Keep in mind that having a PhD doesn't mean you have to go on in life as a bench scientist or work your way up the ranks of academia. There are many different career paths you can follow, from jobs in policy making, journalism, communication, teaching, and consulting. The list is practically endless. So, if it's just a matter of having decided that you don't really want to be a scientist after all, it may still be worth getting your PhD and making use of the degree (and the title of 'doctor' that goes along with it) in another job area altogether. You may be surprised how much value your PhD is, even outside the world of science.

Saving an Old Master Painting: A Set of Experiments Fails, and the Team Faces a Major Setback

For several months, Yousef and Isabel have been making measurements on a paint sample using the SIMS technique. Isabel has also been performing other types of spectroscopy, including IR and UV spectroscopy. When they begin to analyse their raw data, however, both discover that the results they'd been hoping for have not materialised. This represents a major setback for the team, and both of them need to rethink their approach. Even Peter is suffering from this delay as he will not be able to make any further progress with the restoration until he learns more about the composition of the original paint and its degradation products. Isabel takes the setback personally and feels that the poor results are due to her own failings in the lab. Yousef is more philosophical about this downturn in events. 'You win some, you lose some' is his take on the situation. It's best to get back in the lab and just get on with it. Isabel starts to have some serious doubts about her own abilities and is not sure that spending more time in the lab will fix the problem. Rather than jump back into things, she decides to take a couple of days away from the lab bench. She spends time with friends and engaging in fun activities to help recharge her batteries. Her friends convince her that setbacks are common and that she should not take them personally. When she's ready to get back in the lab, she learns that Yousef has drawn up an action plan for how to proceed. Peter has dealt with the situation by doing more reading on the chemistry of paintings. Even though he's not doing any laboratory experiments himself, it helps him to feel more a part of the project. It also makes it easier for him to talk to Yousef and Isabel about their frustrations in the lab.

8

Mentors, Leadership, and Community

If your actions inspire others to dream more, learn more, do more, and become more, you are a leader.
—John Quincy Adams

One of the best things you can do at the start of your scientific career is to find a mentor. A wise and caring guide can mean the difference between wandering around aimlessly in the fog and striding purposefully down the path of academic life and beyond. But don't you already have a mentor, you may wonder? Won't your research advisor play that role? Perhaps, but mentors and advisors aren't usually the same thing. For one thing, an advisor directs, whereas a mentor, guides.

If your research advisor is a natural mentor and is willing to take on that role in your life—and that relationship works for you—count yourself lucky. Not every graduate student is fortunate to have such readily available guidance and counsel from a more senior person. Chances are you'll need to look beyond your lab to find a good mentor. So, what should you look for, whom should you ask, and how can you help your advisor—and yourself—to develop good mentoring skills?

Mining for Gold: Defining Mentorship

Before you start looking around, you'll first need to take stock of what a good mentor is and what you hope to get out of the relationship. A good

© The Author(s), under exclusive license to Springer Nature
Switzerland AG 2022
P. Gosling and B. Noordam, *Mastering Your PhD*,
https://doi.org/10.1007/978-3-031-11417-5_8

mentor has many characteristics, but he or she must first and foremost care about your professional development and have an interest in guiding younger scientists as they move through their careers.

This sounds time-consuming, and it can be. Why would anyone want to take time out of a busy schedule to mentor you? But it's not all about 'taking' on your part. Many good mentors cherish the role of guiding younger colleagues. They gain something by giving back to the community of professionals from which they themselves were nurtured. Now that they've moved up in their careers, these scientists believe it's time to help others make the trek to the summit.

Mentorship is all about experience and wisdom. So, it goes without saying that a good mentor will be someone who is further along on the career path than you are. Before approaching another person and asking them to act as your mentor, however, you need to think carefully about the kind of person and professional you wish to emulate. On a more specific level, is there someone whose career choices you admire? Who has a great work/life balance or is particularly good at getting their work published in top-tier journals?

Take a look at the personality types discussed in Chap. 9 to see how you might benefit from a mentor whose personality type is different than yours.

Importantly, a good mentor should have no ulterior motive in helping you (beyond the intrinsic satisfaction that mentorship provides). He or she should be able to help you meet your own goals (not follow their own agenda) by providing you with support and guidance, modelling successful behaviour, introducing you to a strong network, and helping you identify your strengths and weaknesses as a scientist and a person.

Choosing a Mentor

When choosing a mentor, you'll need to be honest about your own needs and what you think a mentor can do for you. Do you want your mentor to offer you regular advice on how to negotiate graduate school and your career beyond? How specific or general do you want this advice to be, and how much of a time commitment will you require? Do you want your mentor to offer you detailed career and networking advice? Or are you just looking for someone who is a good listener and can act as a sounding board when you find yourself on shaky ground?

If your research advisor is also your mentor, you may want to establish clear goals for your relationship as both a PhD student and a mentee. For example, you may want to meet on a regular basis just to discuss issues outside your research. A good, comfortable relationship with your advisor, as well as a certain amount of personal chemistry, will be key for the mentor/mentee relationship to flourish.

But what if your research advisor isn't able or willing to act as your mentor? If you find yourself in this situation, you'll need to take the initiative and look for someone else. The first place to start is your own lab. How about a Postdoc or even a fellow PhD candidate who has more experience than you? If no one in your lab is a suitable candidate, someone else in your department may be. Some institutes even have a mentor programme in place for those who are unable to find a mentor for themselves. Even if such a programme is available, however, you'll still have to do some work. Mentor/mentee relationships are largely personal, so it's important to have a mentor for whom you have great respect and a warm personal regard.

If you do look outside your lab, be sensitive to possible rivalries or politics between research groups. Even within the same institution, many lab heads are in competition with each other for funding, lab space, and equipment. You won't want to risk angering your advisor by seeking guidance from a direct competitor. The same is true if you consider possible mentors in your field at other institutions; even if you're collaborating with them on some projects, they could still be viewed as a competing lab.

When you've identified one or two individuals who could act as your mentor, it's up to you to approach them. Some people may feel flattered that you've asked for their guidance. Others will turn you down out of fear that mentoring will take too much time, or that you will become overly dependent on them for all your decisions. Don't be hurt if your preferred mentor graciously declines. It's most likely not personal, so remain cordial and move on to someone else suitable.

Once someone agrees to be your mentor, hold up your end of the relationship by respecting your mentor's time and professional responsibilities. You and your mentor should decide together how to move forward and how much interaction you will have. Perhaps you'll meet over lunch once a month or touch base regularly via e-mail, or your mentor will be available whenever you have a specific issue. Whatever you decide, remember that your mentor's role is to provide you with professional guidance and to help you develop independence, not to hold your hand every step of the way.

Working with What You've Got

What do you do if you try all these things but fail to find a suitable mentor? You may want to take a second look at your supervisor. Even if he or she seems less than willing, think of ways you can help your supervisor become a (better) mentor. Start by making an appointment to talk about your needs. Recognise that time is in short supply and make it clear that you don't intend to add to their overly long to-do list. But be up front about your needs. Is it regular discussions you're after, an open-door policy, or just open lines of communication so you feel you can go to your supervisor when you need a bit of guidance and support?

Encourage your supervisor to involve you in group meetings and discussions, and state that you are willing to do whatever extra things need to be done to learn and grow in your field. Volunteer to give a presentation to the department or offer to spend time with a visiting scientist as a way to expand your network. When it comes time to write your first paper, offer to write the first draft on your own and meet with your supervisor for comments and suggestions.

Develop a Community of Peers—Or Become a Mentor Yourself

Professional success doesn't begin and end with having a mentor. Your time in graduate school is an excellent chance to strengthen your professional and social networks and create a community of your peers. Some of these professional relationships may even develop into lifelong friendships and be a source of support throughout your professional life.

Be a leader among your peers. Participate in group meetings and encourage quieter members to speak up. If you don't already have one, start a journal club in your group and invite others in your department to join. Organise social events or team-building activities to help strengthen your relationships outside the lab.

As you move up the lab's food chain, become a mentor yourself by offering to supervise an undergraduate's research project. Offer to teach when possible or provide tutoring sessions for undergraduates interested in pursuing an advanced degree.

As you progress through your career, you'll find that the mentoring you received as a graduate student and Post-doc, and the networks you developed as a young scientist will provide both a firm foundation and a strong scaffolding for your career to grow. When the time comes for you to mentor others just starting out, use your insights and hard-earned wisdom to give junior colleagues a boost. It's also another way of giving back and saying 'thank you' for the help you received early in your career.

9

How to Get Along with Your Labmates, et al.

I prefer the sciences which exhibit nature on a grand scale, to those that are confined to the minutiae of petty details.
—*Jane Marcet*

Working towards a doctorate requires collaboration with others. Your supervisor, a Post-doc in your group, fellow PhD students, the lab assistant, the computer expert, and the technician from the workshop—they will all contribute to your thesis in one way or another. No man or woman is an island, so you'll have to deal with the people around you—whether you happen to like them or not! With some of them you'll have a natural fit, and the collaboration will be pleasant and productive from the start. Other people, well…let's just say they may be harder to deal with. They may have different opinions about planning, be more outspoken than you are (or less), stick too much to the details, or are unimaginative when looking at the big picture. With the aid of a simple tool, this chapter aims to help you to understand others (and yourself) better, how they operate and what makes them tick. Then we talk about how your work can benefit from the inherent differences among people. Creating a pleasant and productive environment for collaboration with both your natural friends, and those you are more or less obliged to get along with, will make your life as a graduate student that much easier.

© The Author(s), under exclusive license to Springer Nature
Switzerland AG 2022
P. Gosling and B. Noordam, *Mastering Your PhD*,
https://doi.org/10.1007/978-3-031-11417-5_9

How to Get the Help You Need from Others on Your Team

To make your thesis research a success you need assistance and cooperation from others. If you assume this support will happen automatically, you might end up disappointed. Not everybody will line up at your doorstep to help you. Despite all your efforts, it may seem that some people are incapable of being any help to you because their way of doing things is so different from yours. Usually we blame these differences, or frictions in temperament, on people having different characters. Often, we just accept the lack of rapport and cooperation (and often blame the other person in the process—after all, it's never our fault!) and look elsewhere for help, or we try to do everything by ourselves. By working alone, however, your progress will slow, and the final result of your efforts might be of lower quality. This chapter describes how people—each with his or her own unique character—can be the key to your success, if you learn the secrets of communicating and working well together. Once you understand what drives others (and yourself) and accept that we are all unique individuals with strengths and weaknesses, your collaborations can be extremely rewarding. In fact, the progress of your PhD research will benefit more when you collaborate with others who differ from you in many respects, than with those with whom you have a natural affinity and share similar strengths and weaknesses.

You, Me, Everybody

There are some people you just can't seem to get along with, no matter what you do, because they appear to have a whole different agenda—maybe even an axe to grind. For some reason they're not interested in fostering a harmonious relationship. This chapter is not about them, however (see more about dealing with difficult people in Chap. 10). Coping effectively with true problem characters is outside the scope of this book. More often, however, some of your labmates will be different enough from you to cause a bit of friction. While they may be just as willing as you are to make their project a success, perhaps they go about things in a different way. These differences might cause so many problems that you're unable to function effectively as a team, and the progress you expected fails to materialise. But no matter how hard you try, chances are slim that you'll ever be able to change the behaviour of others. It is equally unlikely that you'll be able to modify or rein in your own significant character and personality traits. What you can do however, is be aware of what motivates other people (and yourself) and to respect the different ways people approach the same problem.

So what is it that drives individuals to act differently in a team? Each of us is unique, but certain aspects of our character make it easier to feel more comfortable with some people than with others. In order to make sense of human behaviour, psychologists have attempted to categorise these key aspects of character in a variety of ways. For practical purposes we have chosen to illustrate the categorisation of team members' personalities using the four personality types described by Myers and Briggs in the 1950s. Following this somewhat oversimplified—but practical—scheme, we can gain greater insight into others and ourselves. In the following sections we discuss the different characteristics of personality types. Next, we talk about how to identify which type you are, and which type can be assigned to those around you. Finally, and most importantly, we discuss how you can get along with different personality types and even benefit from the fact that other people are different from ourselves! Armed with an awareness of personality types, you can build on your complementary skills rather than trying to turn everybody into a carbon copy of yourself.

MBTI: Getting at the Heart of Personality

In 1920, the psychologist CG Jung made the observation that people are fundamentally different and suggested that, for all practical purposes, we can be categorised by 'function types'. In the forties and fifties Jung's ideas were further refined by Isabel Briggs Myers and Katharine Cook Briggs. They postulated that four qualifiers determine the basis of human behaviour and called it the Myers-Briggs Type Indicator or MBTI. In the course of our lives these qualifiers do not fundamentally change. A component of this theory is that people are essentially different and will behave differently when operating in a team (even when a 'team' consists of two people). Based on this concept, this is where you, as a PhD student, come into play. In order to make your PhD project a success, you will need the help of others. Whether you like it or not, you will have to work with others who are very different than yourself. Understanding the various ways people behave is the first step in working productively together. As discussed in some detail in this chapter, Myers and Briggs devised the following categories for describing character traits: individuals are either (1) introverts (**I**) or extroverts (**E**); (2) driven by intuition (**N**) or sensation (**S**); (3) a thinker (**T**) or a feeler (**F**); and (4) you are either a planner who likes to draw conclusions (**Judge**) or you're more comfortable in a chaotic environment and like to keep things open (**Perceiver**). Altogether there are $2 \times 2 \times 2 \times 2$ combinations possible, hence 16 Myer-Briggs types. Scientists are often (but not always!) found in the NT subclass of this scheme. For further details (and to determine your type by taking a test), wander over to the Myers and Briggs Foundation website.

How You Get Energised: Extrovert Versus Introvert

The definition of extrovert (E) and introvert (I) is slightly different than the one you might be familiar with. In terms of MBTI classification, the key issue in characterising E's versus I's is where they get their energy from. At the end of a long day, do you get energised by attending a social event (E), or do you recharge best by having some quiet time to yourself (I)?

When meeting with others, we all alternate between (inter-)actions and reflections. We talk, listen and think, and talk again. Some of us (E's) have a tendency to act first, and then reflect, followed by action again. While others (I's) start with reflection before they go into action. So E's talk, think, and talk, whereas I's think, talk, and think again.

How You Think: Intuition Versus Sensation

The way you think about things is split into two innate traits: sensation (S)-individuals are fact based. They recall facts from the past, rely on facts in the present and want to know the facts for the future. This is in stark contrast with those with the iNtuition trait (denoted as N, as the I-label is assigned to 'introvert'). N's recall the past in terms of patterns and dream of exploring the future with all its possibilities. The details of the present interest them very little. While sensation-driven individuals look for practical solutions based on past experience, those who are driven by intuition are more interested in exploring new ways of getting towards their goal. Intuitive thinkers are not bothered by the understanding that their thoughts are based on assumptions rather than facts.

Are Your Decisions Driven by Objective Arguments or Feelings?

When it comes to making a decision, we tend to combine factual arguments and value judgements. Thinkers (T's) are more likely to choose or make decisions based on impersonal information. The logic behind the decision is more important than the impact the decision might have on others; for T's, conflicts feel natural. In contrast, feelers (F's) instinctively take into account the impact a decision might have on others. Factual (and unfeeling) data have little influence on their decision-making process. Avoiding conflicts is an important motivator in an F's decision-making process. While MBTI traits are not gender specific in general the F versus T tendency is the sole exception. Slightly more women are Feelers than Thinkers, whereas for men the opposite holds true.

Chaotic Team Members Versus Planners

The fourth and last category in the MBTI classification scheme concerns the way a person organises his or her actions. Some of us like to plan our actions in terms of tasks and targets; those individuals are classified as having judging (J) characters. Others prefer to multitask and plan along the way. Those characters are known as perceivers (P) and enjoy the chaotic process in which the final goal only becomes clear towards the end. While J's dislike stress and try to avoid it by planning their activities, P's work best under a certain amount of time pressure and get energised once the deadline is approaching. Results at the midpoint are most easily measured for the J's. They tend to have finished half of their tasks according to plan. Whereas may P's have gathered lots of information and are working on multiple tasks, it's difficult to see how far they have come in reaching their targets.

Which Type Are You?

Now for the fun part. In order to benefit from the MBTI classification scheme you'll first need to know your own type. The short description given above might give you an initial hint. To make a more thorough check, you can also complete the official MBTI® instrument questionnaire available on the Myers and Briggs Foundation website.

How to Collaborate with Your Personality-Type Counterpart

Although the MBTI is meant to classify individuals into one of the sixteen categories, with each category or type distinct from the fifteen other types, we'll make a little shortcut here. For the four classifications we'll restrict ourselves to discussing the differences between team members with opposite personality types: E versus I, S versus N, T versus F, and J versus P. Once you have a sense of the personality differences between yourself and others, our suggestions on how to get along best with each different type will help improve your working relationship with—and understanding of—a range of different people.

Extroverts Versus Introverts

The various ways that people think and act will already be apparent during your first group meeting. As soon as some new issue is put on the table, the E's in the team will immediately discussing it. They develop and express their opinions while the discussion is going on. Whereas the I's in the team need to think first before they're ready to discuss the issue. These differences could lead to mutual irritation within the group. The extrovert team members might be annoyed that the introvert team members appear to be uninterested, since they're not yet participating in the discussion. At the same time, the I's may be irritated by the E's because they start shouting out all kinds of ideas even before they've given them any thought. To resolve this difference in style, each type needs to respect the others' approach. So, the I's need to let the E's talk, as it's their way of developing an opinion. Moreover, what the E's say at the beginning of the meeting may not reflect their final point of view. Give the E's a chance to change their minds. At the same time, E's should ensure that the I's get the chance to contribute their opinions after they've had time for some initial reflection on the new issue. If the leader of the discussion is an extrovert, she/he might overlook the more introverted team members. Since I's might have a hard time breaking in during the meeting, particularly if the E's are dominating the discussion, the I's will continue to think and listen. Eventually when the I's have a nearly complete picture for their input, they'll start talking. This style of communicating is often less understood or appreciated than it should be. The others may feel that the I's should have put their ideas on the table earlier in the discussion. To the E's, it might even come across as 'arrogant' to not participate in the initial brainstorming, then suggest a definitive solution all at once. Both E's and I's should make sure that the introverts on the team get involved early on during the meeting. By nature, introverts will typically hold back at the beginning of a meeting, but eventually they will make their opinions known. During the break, halfway through the meeting, everyone will need to get energised again. The I's are going to sit alone with their coffee, but that doesn't mean they are not interested; they just need to charge their batteries in isolation or perhaps with one other person. The E's are energised by the outside world and will usually talk in groups during the break. Whichever type you are, be respectful of your personality counterpart and try to see the world through their eyes.

Intuition and Sensation Are Both Necessary for Success

Misunderstandings and poor communication between iNtuitive thinkers and more Sensation-orientated individuals are often sources of conflict in teams and can result in a lack of progress or even a stalemate when it comes to decision-making. This situation can be particularly fraught as most projects need both personality types to be a success. The absence of sensation-type characters on a project can still yield good results in terms of the big picture, but the project will lack the necessary scientific data or facts on which the picture should be based. On the other hand, a team of only sensation-types might fail to discuss what the data are actually good for or what they're out to prove in the first place. The reason sensation- and intuition-type personalities have difficulties working together is quite simple. The dreams and schemes of potential future projects of intuition-driven individuals are constantly being interrupted by factual information from those who are sensation-driven. Such facts may 'prove' that the intuitive thinkers' conception of the big picture is impossible. On the other hand, the carefully gathered facts, based on scientific findings, by the sensation-type scientist are downplayed by the intuition-driven researcher who considers them to be just a bunch of details.

The bottom-up approach from the sensation researcher would benefit quite a bit from the global top-down picture of teammates with intuitive minds. At the same time, the facts gathered by S types can be essential for verifying or falsifying the initial hypothesis from the N-type team members. We can't state this often enough: respect and understanding for individuals with the opposite personality-type is key to avoiding conflicts in a team. The key to success is an awareness that the best solution to a problem can only be achieved if team members with opposite personality types work well together.

Feeling is More Important in Science than Thinkers Want to Believe

Science is based on facts, so facts are an essential part of the scientific process. It's probably a safe assumption that Thinkers are more likely to be attracted to this type of work. However, as has been argued before, to make progress in science, teamwork is needed. To get teams working well together, many compromises will have to be made. Such solutions are not only based on the average point of view but will be affected by personal relations and perspectives. A Feeler can play an important role in making sure that the team works in harmony.

Science can seem very objective at first glance: results are results and should not depend on the way you think about them. The conclusions you make are fact based. However, the impact your conclusions have on the scientific community can very much depend on your presentation. Presenting conclusions is quite a subtle thing, in particular if you want to get some recognition for your work. If you are too modest, the scientific community will not notice your contribution, and you will get little credit. If you present your findings with too much enthusiasm, for example, or down-play the work of others in the field, you might easily make enemies. These people, in turn, probably won't credit you for your contributions. Thus, having the attributes of a Feeler can make a substantial difference in how successful you are in your work, no matter how objective and analytical you think a successful scientist should be.

Judgers and Perceivers Play Crucial Roles at Different Stages of the Project

Judgers and Perceivers will use different approaches in planning a new project. They might get irritated by the approach of their counterpart, but in a respectful collaboration a mixed team has an enormous advantage over a team with only P's or J's. Before they even start on a project, Judgers will want to make a plan in which the goals are defined and the routes towards that goal are outlined. For instance, the countdown plan described in Chap. 19 will probably appeal more to Judgers than to Perceivers. Typically, Judgers are restless at the beginning of a project when the plan is not ready. Once they know how they want to execute the plan, J's will start feeling more relaxed. For Perceivers, the opposite is true. In the beginning P's start gathering information, seemingly without a plan. They are typically quite flexible and relaxed. Towards the end of the project, when J's have already finished 90% of the tasks and move on towards the last bit, the Perceivers are just becoming very active as they wrap up all the loose ends.

These differences in approach can easily be a source of conflicts. Perceivers might get irritated by Judgers jumping in early to force the outcome (i.e., by making a rigid plan), while Judgers might feel that Perceivers have accomplished very little at the halfway mark, since they have not produced tangible results. Perceivers can benefit from the structure that a Judger brings to a project in the beginning. However, if the final goal is changed as the deadline is approaching, Judgers tend to panic: all their hard-won results may have lost their value. At that point, Perceivers are at their best; their adaptive and flexible way of working allows a reorientation of the plan, even as the deadline is approaching.

More formal team meetings tend to have an agenda. Such an agenda is very important to J's, while P's pay little attention to any agenda. If the chair of a meeting has a Perceiving character, he or she might forget to stick to the agenda, or even to make one. This makes Judgers quite nervous, and unproductive. A 'J' has a hard time starting on any project or meeting if there is no plan. In contrast, if the chair of the meeting is a Judger they might stick exactly to that plan, while the contributions of the Perceivers, who prefer the flexibility to 'wander around a bit', might fit less well into a rigidly structured scheme.

A Varied Mix Makes a Good Team

In the sections above we discussed individual aspects of personality and how they interact, rather than a combination of traits (known as temperaments) such as SP's and SJ's. We refer the reader to the literature for this next level of MBTI classification. The discussion of the role of opposite characters in a team is meant to illustrate the concept that team members with different inclinations contribute in different ways to the final result. Knowing your own strengths (and weaknesses), as well as those of others, is the first step in getting the most out of your team. Respecting each other and your unique working and communication styles will lead to a successful team relationship. It may help to know that some organisations charged with setting up professional teams to execute complex tasks make a point of creating teams of individual with different personality types.

The Golden Boy/Golden Girl Syndrome (He—or She—Who Can Do No Wrong)

Okay. So much for personality types. They're a great tool for helping you identify your own working style and that of others (including your supervisor) in your lab. But there is often a situation, more common than many people think, that can cause friction in your research group, and understanding personality types will not do much to help solve it. This is the Golden Boy or Golden Girl Syndrome. In many groups, one of the PhD students (or sometimes a Post-doc), tends to stand out. The Golden Boy or Girl produces data that go straight into high-impact journals. During group meetings, he or she seems to come up with all the bright suggestions. The head of the lab (i.e., your supervisor), praises this individual to the high heavens, and tends to spend all of his time with him or her. This further improves the contributions of the Golden Boy or Girl, leaving the others trailing in their wake. So how to cope? First, you need to accept that we're not all created equal and that the Golden Boy or Girl may truly be exceptionally talented. Much of the attention this individual receives, however, may be less due to talent and more to the force of their personality. There is little point in fighting this or trying to get an equal amount of praise and attention for yourself. This strategy will most likely backfire. However, since you do deserve a fair share of your supervisor's undivided attention, make a point of asking for regular meetings to discuss your work. These could occur every month after you've filled in the Monthly Progress Monitor and have concrete issues to discuss. During your meetings with your supervisor, make clear to him or her that you appreciate their help. Make clear that you're making progress in a particular area. Don't mention the Golden Boy or Girl, and certainly don't try to complain about this individual or put them down. On the other hand, don't allow this individual to steal the spotlight and the credit all the time. You deserve respect from your supervisor, as well as the other members of your group. So stick up for yourself and make sure you get your share of the limelight.

Saving an Old Master Painting: The Team Learns How to Overcome Their Different Styles of Communication and Work Effectively Together

Early on in their collaboration, Isabel, Yousef, and Peter realise they are a special kind of team. They do not belong to the same research group and each has a different supervisor. They've been brought together by their common research goal, and the ability to work together in a cooperative way will be key to their success. Isabel points out that they should understand and respect each others' different working habits and approaches to solving scientific problems. Otherwise, their fragile collaboration, without a formal structure, could easily fall apart due to conflicts. Yousef agrees with Isabel, and suggests they use a personality model to understand their behaviour and internal driving forces. So, they decide to take the MBTI test and discuss the outcomes.

Of the three, Isabel is the least interested in taking such a test to characterise their personalities, but she went ahead anyway, because she likes the idea of discussing together their behaviour as a team. The test indicated that **Isabel** is an **ESFP**, an extrovert, sensing, and feeling perceiver. Initially, Isabel disagreed with being labelled as a Perceiver (score 60% P, 40% J). She argued that she organised her office properly, which is a typical trait of a Judger. On second thought, after thinking about her private lifestyle, she realised that organisational skills did not come naturally. Isabel found it an effort to organise everything in just the right way in order to conduct complex experiments properly. Ultimately, she recognised herself in the description of ESFPs: They are open and enthusiastic towards the world. They seek the company of others and have a deep concern for friends.

Yousef, who introduced the MBTI test to the team, enjoyed taking it. According to the test results **Yousef** is an **ENTJ**, an extrovert, intuitive and thinking judger. Being classified as an extrovert surprised him somewhat, since he did not see himself as having an extremely outgoing personality. However, In terms of acting first and then thinking (E), versus thinking first and then acting (I), he realised he was definitely an E. Yousef recognised himself quite a bit in the description of ENTJs: he enjoyed being in charge and managing his projects with conceptual models. For instance, the goal-setting strategy used in the monthly progress review appealed to him.

Peter was curious to find out why some things seemed to work very well in their team, while others failed. He wanted to fill out the questionnaire alone, however, despite the suggestion by the other two to do the whole exercise together. The MBTI test indicated that **Peter** is one of the rarer types, an **INTP**, who is an introvert, intuitive and thinking perceiver. The thinking part was new to Peter. He'd never realised that his seemingly detached behaviour was described by the thinking type. Peter had no problem with the typical INTP description: as he was driven to conceptually understand phenomena, and in spite of being somewhat detached from other people, he was an excellent teacher, in particular for advanced students.

Once they all knew their MBTI classification Isabel, Yousef, and Peter discussed their differences and how to make the best use of them, including how to cope with the 'think first' (I) attitude of Peter, the 'attention to details' (S) characteristic of Isabel and her open and enthusiastic attitude towards others (F), and Yousef's tendency to organise things up front (J). What they learned most from the whole exercise was that it's essential to the success of their team that they bring different personality traits to the project, and to recognise that they are stronger as a team due to the combination of their individual strengths.

10

Group Dynamics: Dealing with Difficult Colleagues

Man needs his difficulties because they are necessary to enjoy success.
—Abdul Kalam

In the last chapter, we talked about how to get along with—and work along-side—people with a variety of personality types. In this chapter, we take a closer look at how to cope with labmates and colleagues whose behaviour could be labelled as 'difficult'. By difficult we mean the type of person with a severe 'personality quirk', who can toss major obstacles in your path that create roadblocks to your progress. Why are some people difficult to work with and be around? We'll leave the answer to that thorny question to the psychologists and psychiatrists among us. In the meantime, if you've got a difficult individual in your midst, you'll need clear strategies to minimise any potential damage they can cause to your progress and your career.

If you learn how to cope with contrary colleagues early in your career, particularly in the competitive atmosphere of a lab, you'll develop valuable coping and people-management skills that will serve you time and again, wherever your career path takes you.

In an ideal world, your lab would contain only bright, capable people working harmoniously together in the pursuit of scientific knowledge. If this describes your lab, count yourself lucky. Most scientists work in close quarters with at least one person who tries your patience or is difficult to get along with.

© The Author(s), under exclusive license to Springer Nature Switzerland AG 2022
P. Gosling and B. Noordam, *Mastering Your PhD*,
https://doi.org/10.1007/978-3-031-11417-5_10

Take a look around your lab or department (and in a mirror, too!) to see if you recognise any of these 'types' who have the potential to sink your career—or your self-esteem:

- **Star Researcher (a.k.a. The Hotshot)**: The Star Researcher is on the fast track to success—or so they think—and has an ego to match their ambition. He or she dominates group meetings and touts their own successes while belittling the contributions of others. Your supervisor gives them the best projects and showers them with attention and praise.

- **Energiser Bunny**: This dynamo seems to live in the lab. He or she is there when you arrive in the morning, and when you leave at night, and seems to run twice as many experiments as anyone else. All this would be fine if it weren't for their tendency to treat with derision anyone who doesn't show the same fierce dedication that they do.

- **Stealth Bomber**: The Stealth Bomber attacks without warning. Right in the middle of a group meeting or department gathering, he or she will say something about your latest failed experiment or cock-up in the lab. The Stealth Bomber operates best in front of an audience and loves nothing more than to ambush others.

- **Know-It-All**: Without any prompting, this person will launch into a lecture on the right way to do a procedure or protocol. They'll look over your shoulder and announce that what you're doing is *all wrong*. 'Here, let me show you how it's done' is the Know-It-All's mantra as they pluck a pipette from your hand.

- **Woe Is Me**: Ah, the chronic complainer. Everything in this person's life is grist for the mill. Experiments aren't going well, she/he isn't getting along with their supervisor, there are problems in their personal life, and the lab equipment is not up to par. If it exists, this individual will complain about it.

- **Hornet**: A prime candidate for anger-management coaching, the Hornet will explode with wrath for no reason at all—or if confronted, challenged, or rubbed the wrong way. You and everyone else in the lab walk on eggshells in fear that the Hornet will deliver a nasty sting.

- **Sneak Thief**: The Sneak Thief borrows your equipment and expertise and picks your brain for ideas, then refuses to give credit when credit is due. When the Sneak Thief has a success, they'll say they did it all on their own.

- **Who, Me?**: This person has a hard time keeping commitments. Suppose you've decided to work on a project together and have divided up the work. But when it comes time to deliver you might hear something like this: 'Who, me? Was I supposed to do that experiment? Order those supplies? Calibrate the machine?'.

Strategies for Coping

Perhaps you're the type of person who usually turns the other cheek in difficult situations and prefers to avoid conflict at all cost. If that's the case, your strategy so far has been to ignore the difficult person. But in a lab situation, avoiding the difficult person will only make matters worse: He or she will go on being difficult, and you'll start to feel increasing discomfort, not to mention resentment. Bringing the problem to your supervisor's attention may seem like another option, but not all supervisors are good managers and may not know how to handle this situation effectively.

So, how do you deal with a difficult co-worker? Each type of person requires a different approach, but there are some simple things you can do to defuse the tension. For some types of difficult behaviour, the best approach may be to talk to the individual about how their behaviour affects you. For other types, more subtle and oblique ways of dealing with the problem behaviour may be required.

With the Star Researcher, for example, it's easy to get defensive: 'Why does *she* get all the attention?' In this case, the best response is no response. Confrontation may cause things to escalate, and you'll end up with a powerful foe. When alone with your supervisor, resist the temptation to mention your irritation with the Star Researcher's ego; criticism from you will seem like sour grapes. Second, concentrate on producing great work. When you submit your own (dazzling) work for publication, the peer-reviewers won't know or care about the Star Researcher's outsized ego. It's the work that counts in the end, so make sure yours is top-notch.

To the Stealth Bomber you might say: 'During group meetings, I've noticed you habitually bring up problems I'm having with my research. I understand that this is fodder for dynamic discussions, but I'd feel better if I could bring up those issues myself.' The advantage of this approach is that by explaining *why* a certain behaviour upsets you, you focus on the behaviour rather than the individual. By being direct, but subtle, you also allow the Stealth Bomber to save face by (hopefully) getting them to see your point of view. This approach also lets the Stealth Bomber know you're aware of what they're doing. Every time it happens, bring it up again until they stop.

With the Woe-Is-Me chronic complainer, you might try adopting a stance of neutral listening rather than joining in and feeding the complaint cycle. For example, acknowledge what the complainer is saying by nodding and making neutral statements such as, 'Hmm, I'm sorry to hear that.' Let the complainer moan about how bad everything is for two minutes and then move into problem-solving mode. You might say: 'It must not be easy to get work done when your equipment keeps breaking down. So what are you going to do about it?' In short, reward positive action, not endless complaining.

When dealing with aggressive individuals such as the Hornet, the best way to cope with an angry outburst is to do nothing. In some cases, it might be better to let such an individual rant. Remain cool and detached, and when they're finished, walk away. Or, depending on how volatile the situation is, you might suggest that you'll discuss the issue when they're ready to talk calmly about it. By adopting a Zen approach and refusing to allow an outburst to escalate, you'll eventually stop being a target of their anger.

The Know-It-All can be particularly irritating in the competitive atmosphere of a lab, where everyone is working hard to become an expert. One way to defuse the Know-It-All is by agreeing with everything they say. Nod thoughtfully and then introduce your own thoughts and opinions in a questioning manner: 'Your way of doing that procedure sounds good, but have your ever considered…?'

As for the Energiser Bunny, so what if he/she puts in 16-hour days in the lab and runs marathons on the weekends? If that's not your style, so be it. Embrace your positive qualities and don't beat yourself up because you work at a pace different from someone else's. It might help to find subtle ways to let the Energiser Bunny know that your work is just as important to you as theirs is to them. If you make it clear that you won't be intimidated by their input—or prodigious output—you may even earn their respect.

Monitor Your Response

Finally, take an honest look at how you react when dealing with a difficult person. Do you get defensive, angry, intimidated, irritated? Or are you able to brush it off? A key aspect of successfully dealing with difficult people is having confidence in your own work. Building confidence takes time, but as you start to amass a steady stream of successful experiments and publications, other people's attitudes and behaviours will matter less. In the short term, it might help to remember that difficult people often act as they do out of fear or poor self-esteem. And ultimately, because you can't really change another person's behaviour, all you can do is change how you handle it. So keep working on your own goals and don't allow others to undermine you.

Working with difficult people is never easy. But if you learn how to cope with contrary colleagues early in your career, particularly in the competitive atmosphere of a lab, you'll develop valuable coping and people-management skills that will serve you time and again, wherever your career path takes you.

11

The Art of Good Communication

I was taught that the way of progress was neither swift nor easy.
—Marie Curie

Is poor communication with your supervisor getting in the way of your progress in the lab? Perhaps you've reached an impasse in your research and can't see a way through. Or maybe it seems that—from your supervisor's point of view—nothing you do is good enough. If you and your supervisor have different expectations of your output, and the two of you haven't spoken in months, then a lack of communication is surely holding you back.

Once settled into their projects, many graduate students are left to work things out on their own. That's as it should be—to a certain extent—as much of graduate training is focused on the ability to meet problems head on and solve them on your own or with input from your team. But your supervisor must ultimately approve your thesis, so keeping the lines of communication open is crucial. Don't wait until you get into serious difficulty before knocking on your supervisor's door. Even if your supervisor keeps their distance, as a seasoned researcher they should be able to provide appropriate guidance, and—one hopes—a neutral perspective. If you get the sense that your supervisor is putting his or her interests above your own, initiating communication on a regular basis will give you an opportunity to voice your concerns.

Some people are born communicators; if you're not one of them, however, and talking to your supervisor feels like talking to a wall, take heart: good communication skills can be learned. If you're having trouble connecting

© The Author(s), under exclusive license to Springer Nature Switzerland AG 2022
P. Gosling and B. Noordam, *Mastering Your PhD*,
https://doi.org/10.1007/978-3-031-11417-5_11

with your supervisor in a satisfying way, the key to better communication is understanding your supervisor's personality and communication style, as well as your own. Everyone is different. Some prefer the free-and-easy approach, while others like more structure. Either way, better communication involves planning, practice, and a conscious effort on your part. If communication with your supervisor is poor or non-existent, and has been from the beginning, don't blame yourself. It's much better to find a way around the impasse. It's also not a good idea to try to change your supervisor's ways; it won't work. Instead, focus on what *you* can do to improve the situation.

Understanding Different Styles of Communication

Does your supervisor always seem to address the lab as a whole, rather than each of you as individuals with different needs, skills, and abilities? Perhaps in your weekly group meeting, she scans the room and asks, 'Everything going okay? Any problems? No? Great,' before dashing back to her office or to another meeting. This kind of behaviour doesn't make your supervisor a bad person; it may mean she is busy and possibly insensitive to cues from lab members about the need for regular contact.

Perhaps your supervisor talks to you on an individual basis, but he's a 'hit and run' artist, tossing out a query about your progress as he breezes through the lab and then hides behind a stack of journal articles on his desk.

Take a look at Chap. 9 for insights on different personality types and communication styles.

If your supervisor is an assistant professor just starting out, she may spend most of her time in the lab working beside you. If that's the case, there will be many opportunities for discussions, both formal and informal. And unless your supervisor is very bad at communicating, good rapport should develop naturally.

If your supervisor is established at the institution and highly regarded in her field, she may rarely appear in the lab at all. In between international conferences, she might stick her head in the door for a quick hello and only meet with her most senior Post-doc to assess the lab's progress. If this is your situation and you feel like a 'worker bee,' with a supervisor who is remote or hard to approach, it can be difficult to speak up and make your concerns known.

But whatever your supervisor's style may be, you can always find a way to make yourself heard. The most valuable thing you can do is to make an appointment to talk face-to-face whenever you have something important to discuss. Even if you have lots of access to your supervisor and engage in many informal chats, a formal talk will allow you to structure your questions and clarify important issues. If you prepare well for the meeting, and stick to your agenda, all you'll need is 15 minutes or so of your supervisor's time.

No matter how busy your supervisor is, plan to meet at least once a month—more often is even better—to discuss your research and other issues you'd like to address. Suggest a time of day when a meeting is likely to be most successful. Is he more focused first thing in the morning? Then make your appointment before he's swamped with other priorities. Immediately after lunch is another good time. Avoid making an appointmens late in the day, however, as it's likely to be cancelled when other priorities press in and the end of the workday approaches.

Structured Communication is Key

Informal, spontaneous communication plays an important role in building relationships and establishing trust. Informal chats about work or other common interests can help build rapport, and the more comfortable you and your supervisor are with each other, the better. A good rapport based on trust and mutual respect can be a great asset.

A warm relationship, however, is not something you can force, but you can still make progress without this kind of rapport. The most crucial form of communication will likely take place during regular, short, face-to-face meetings between just you and your supervisor.

Once your appointment has been set up, take time to prepare. As a general guide, we suggest your preparation time equals at least the amount of time you plan to meet with your supervisor. Go to your meeting with a written list of questions and concerns. Keep them brief—no more than three issues per meeting. Be specific; it won't do any good to ask, 'So, how do you think I'm progressing?' A question like that will just encourage your supervisor to respond in general terms or say something encouraging but meaningless, or—worse—disparaging but meaningless. If you need guidance on how to move your research forward, for example, come to your meeting with two or three of your own ideas about how to proceed. Give your supervisor enough context to be able to provide you with helpful input. If you haven't spoken for a while, give him a brief summary of your most recent results.

During the meeting, take notes and jot down your supervisor's suggestions, assuming it's okay with her; some people find it disconcerting to have their remarks written down. As you chat, gauge your supervisor's enthusiasm and interest by paying attention to body language and other non-verbal cues. At the end of the meeting, thank your supervisor for her time and immediately send a follow-up e-mail that summarises what you discussed. That way, you'll have a record of your questions or concerns and your supervisor's responses. Print out the correspondence (or save it in a file on your computer), along with your original list of 'talking points,' for future reference.

In addition to your face-to-face meetings, you may want to chat with your supervisor whenever the chance arises, as well as send him informal monthly updates of your progress via OneNote or by e-mail. Even if your supervisor is unwilling to work with you on creating a Monthly Progress Monitor, sending an e-mail at the end of every month, with a brief summary of the experiments you've done and results you've achieved, is another effective way of keeping your supervisor up to date on your work.

Take a look at Chap. 6 for details on the Monthly Progress Monitor.

But none of this should substitute for regular, short, structured meetings with an agenda you prepare. Allowing too much time to pass between structured communications may cause your research—and your relationship with your supervisor—to veer off in a direction you don't want it to go.

With all the focus on structure, why bother to have a meeting? Can't it all be done by e-mail? No, not really. E-mail and other electronic forms of communication are useful, but they aren't adequate in this context. Even if you're reciting lists and focusing on facts during your face-to-face meetings, you're sending and receiving a complex set of verbal and non-verbal cues that are crucial to establishing trust, which is the foundation of a strong working relationship. E-mail fails to convey this vital information. Emoticons are no substitute for real emotions. Meeting frequently and regularly with your supervisor, asking relevant questions, and documenting her input will increase the probability that good communication flows in both directions and that your research is in line with what your supervisor wants and expects.

Learning good communication skills in an unstructured environment can be a challenge. But fostering effective communication with your supervisor, particularly if he or she is a poor communicator or difficult to approach, is a skill that will serve you well throughout your career. Even if you become an independent entrepreneur without a boss, you will surely have clients and colleagues who will benefit immensely from your ability to communicate well.

Effective Communication in the Lab: A Practical Approach

Progress in your research is by and large determined by—if not dependent on—interactions with others. Spending some time to solve communication issues can be quite helpful in smoothing the way forward and facilitating your PhD project.

Effective communication will yield more helpful contributions from others and improve the progress (in both time and quality) of your PhD research. Again, once you've learned how to continuously educate yourself in improving key communication skills, you will benefit from these efforts throughout your career.

For example, driven by the desire to stay in our 'comfort zone', we all have a natural tendency to drop by the office of a friend in case we need a favour, whereas requesting help through more formal channels (e.g., by sending an e-mail) might feel like the best approach with those we don't have much rapport with in the lab. While there is nothing wrong with preferring to chat with a friend, we argue here that you should spend even more attention and energy on communicating with those you get along with less well.

So, what are the ingredients of effective communication? In all cases it starts with a sender, let's say you, who wants to get a response from someone else. To that end, a message is sent to the receiver. To make the communication effective, it should be done in such a way that the receiver carries out the desired action as intended by the sender. We all know it's easy to ask for something; however, our request won't necessarily result in the action we were aiming for.

We can divide communications into three categories that differ in the timescale in which the desired action should take place, as well as the method used. First, short-term communications include asking someone during a face-to-face interaction what you want that person to do right now (e.g., can you please take that sample from the fridge?) We've all been trained in this type of communication practically since we were born, although the 'sample' for a young child will be a different item than that for a PhD student.

Second, at the other end of the spectrum, there is long-term communication, in which either the *reception* takes a long time (e.g., it may take months or even years before someone reads an article you have published) or the *action following the reception* takes a long time (e.g., although someone may be listening right now to your oral paper, it may be quite a while before this results in follow-up experiments by the 'receiver' or a mention of your work in a scientific journal). Fortunately, most PhD students will have the opportunity to give oral presentations and/or write articles about the research they have done. With feedback from your supervisor and members of your research group you'll no doubt improve your long-term communication skills along the way.

The third category of communication takes place on an intermediate time scale: you send the message and expect actions from the receiver on a timescale of days. These communications could be sent as chat, e-mail, voicemail, text message, telephone call, or in a face-to-face encounter. We typically spend limited time in selecting the optimal channel for getting the best response from the recipient upon receiving our message. We discuss here four potential channels and how to use them most effectively. In addition, you also might want to take into account how the receiver of the message prefers to be communicated with. Some people prefer e-mail, for example, over the distraction of having others constantly dropping by their workspace.

E-mail: While traditional communication styles have their own form of etiquette, e-mail can often be quite blunt. In the past two decades, e-mail has replaced many other modes of communication. The general acceptance of e-mail has been helped by its ease of use, the possibility of sending attachments with your message, and the comfort of knowing that messages are easily stored on your laptop (thereby allowing both parties to refer to the message later on).

For the sender, e-mail may feel as casual as dropping by someone's office, so senders tend to forget that the message will be read and perceived in a way that is somewhere between a phone call and a proper letter. Poorly phrased wording, typed in a hurry, can be easily misinterpreted by the receiver and could lead to misunderstandings. The message and all the megabytes of the attachment arrive instantaneously in the inbox of the recipient but are not always read straightaway (many people have hundreds of unopened e-mails lurking in their inbox), let alone acted upon in an appropriate and timely manner. So, before you push the 'send' button, we suggest you verify three things:

1. **Ensure the e-mail has an** appropriate subject header that states the actions that are expected from the person (or people) receiving the e-mail. If no action is required, you can start the header with FYI: for your information. Sometimes the action is so clear that there is no need to include any information in the e-mail itself (e.g., 'Tuesday's progress meeting is cancelled'). In that case, you can close the header with: EOM (end-of-message). 'Please review enclosed draft article by next week' is a good example of a header indicating *what the e-mail is about* (the draft of an article), *what needs to be done* (review), and *when the sender hopes to get a response.* In the main text you can politely ask for help and inquire whether it suits the receiver's schedule to do the review within the allocated time frame.

2. **Read the e-mail again**, once it's finished, to check for typos and, more importantly, to view it through the eyes of the recipient to ensure your message will come across as intended. Restoring the damage caused by an ambiguous e-mail will take much more time and effort than spending the extra minute to double check whether your intentions are clear.

3. **Make sure your e-mail message conveys clearly** what you expect from the recipient(s). Do this at the beginning of the e-mail and repeat this in the summary statement that closes your e-mail. In some formal circumstances it may be desirable to have your communications documented. E-mails are perfect for this purpose (demonstrating once more why you need to be careful what you write). Be aware that in certain circumstances you'll need a formal record of your communications, especially if you're far removed from a collaborative setting.

Phone calls, Chat, WhatsApp, Skype, Zoom, Teams: An unexpected side effect of the massive adoption of smartphones in recent years is the burgeoning number of apps for communicating by video chat and calls, such as Skype, WhatsApp, Zoom, Teams, etc. Most—if not all—PhD students use mobile phones for work-related communication. Calling someone on his/her mobile is an easy way to make contact and get a message across. However, there's a fair chance these days that the call will not be answered, and you'll be directed to a voicemail box, which may never be checked. But with all the other options available, your phone is still a launching pad for multiple avenues of communication. With free calling and text messaging services, the world, as the saying goes, is truly your oyster.

Face-to-face: There's nothing like a good old-fashioned face-to-face meeting. Such a meeting can be a scheduled event (in which you sit together formally with your supervisor to discuss your progress) or a quasi-spontaneous encounter at the coffee machine. You may have noticed that some very effective people spend quite a bit of time in such informal encounters. In a face-to-face meeting not only is the message communicated, it's also supported by non-verbal cues from the sender. The actions to be taken by the receiver—to a large extent—typically depend on this non-verbal support. In-person interaction allows the sender to adapt (in real time) to the non-verbal cues and the verbal message while relaying it.

Take a look at Chap. 15 on the pluses and pitfalls of remote collaboration.

Ground Rules for Optimising Effective Communication

Since communication is so important in any job requiring the execution of complex tasks, you have likely set yourself some guiding principles about how and when you use the above-mentioned communication channels, in the event you want to get something done on the timescale of, say, a day or a week. Most likely, however, you don't make a conscious decision every time you communicate information on how you plan to carry out a particular activity or task. Here, we provide a few generic tips that might help you further improve your communication skills.

1. **Spend the most time communicating with those who are the least willing to collaborate.** In any lab—just like in normal life—you'll find you get along better with some people than with others. Communicating with friends feels much easier than with someone who is perhaps not willing to take time out of their life to help you. We all have a natural tendency to spend more time with the people we have a good relationship with, which makes interpersonal communication much more effective. However, the rate of your progress on a particular project is often determined by the task you're having the most trouble with (i.e., the 'weakest link' principle). Hence, it might be a good idea to spend time improving your communication skills with any individuals who are disinclined to provide you with the relevant assistance you need.

2. **Consider the communication channel you choose, given both the message and the recipient**. The type of message you wish to send makes some communication channels better suited than others. If you need to cancel a regular progress meeting, an e-mail will do the job. However, if you have decided that your demanding PhD project does not permit you to spend a lot of time helping others with their own work, it's probably a better idea (though not easier) to communicate this in a face-to-face encounter. The tougher the message, the more inclined we are to use impersonal channels (such as e-mail), even though a personal approach is usually the better option, given the long-term collaboration you're involved in. It may help to remember that getting a PhD takes years, and you will likely work with the same team for most of that time.

3. **Try to make communication a two-way event**. There's a subtle difference between a request for help and a command for support. A command phrased as a polite request might be more effective in the end, even though as a PhD student it's unlikely you'll be in a position to give commands to anyone. Moreover, the recipient is more likely to perceive your desire as a request for help if you set up the communication as a mutual interaction. For instance, in an e-mail you might mention that you're looking forward to their input for solving a technical problem that's slowing the progress of your project. Or you might close a voicemail by saying: 'please let me know if this message is unclear, or you believe there's a better way for me to achieve my goal.'

4. **Combine several communication channels**. As discussed above, all communication channels have their pros and cons. In many cases it might help to combine channels. In some cases, combining e-mail and a phone call can enforce the message, particularly if it's urgent or important. Make a quick call to say that you're crafting an e-mail with additional details, but that you wanted to let the recipient know ahead of time that you're responding to his earlier message.

5. **Verify that the person received your message and acted as intended**. Once you've sent the message (e.g., by e-mail), you might feel your job is done. But you did not send the message just to push the problem onto someone else. You should verify that the intent of the message was understood, and that the recipient will take the appropriate action(s). In some cases, it might be obvious that the action was taken, while in others, it would be a good idea to confirm that the appropriate action was indeed

carried out. One way to be sure, is to let the recipient know exactly what your expectations are.

6. **Ask how you can help turn your request into a successful action**. Although you will likely need help from others with some aspects of your PhD project, simply sending a message or query may not be enough to get the help you're looking for. As mentioned above, communication is a two-way process, so be aware of the type of communications you engage in, in order to get the input you need from others.

In summary, good communication through a variety of channels is a vital part of successful collaboration with others. It's worthwhile, therefore, to invest in building your repertoire of communication skills. Once you master those skills, you'll benefit throughout your career from your increased facility with several modes of communication, as well as getting the desired results and input from others that you need.

12

Mastering Presentations and Group Meetings

The newest computer can merely compound, at speed, the oldest problem in the relations between human beings, and in the end the communicator will be confronted with the old problem of what to say and how to say it.
—Edward R. Murrow

Many people rank public speaking among their greatest fears. But talking in front of an audience doesn't have to be a frightening prospect. The trick is to be prepared, know your stuff, and practice until it feels natural to talk about your work in front of a group.

For several months now, you have been designing experiments, carrying them out, and collecting data. At some point, your advisor will ask you to present your findings to others. Perhaps your first presentation will be in an informal setting, such as other members of your lab during a weekly or monthly group meeting. Or you may be asked to give a presentation to the entire department. While still a graduate student (and count yourself lucky if you're given the chance this early in your career), you may even be invited to present your research at a large regional or international conference (see Chap. 14).

Let's first discuss giving a presentation to a fairly large audience. This type of talk is by nature more formal than a group meeting and requires more structured preparation.

Get Prepared

Although giving a presentation may seem daunting at first, spending time to prepare will take a great deal of *angst* out of the process. Whether you have an audience of two or 200, your approach and general objectives are the same: getting your message across in clear and simple terms that leave your audience hanging on the edge of their seats and hungry for more.

Where to start? Not by rushing to the computer to experiment with fancy PowerPoint templates and a snazzy array of bullet points and arrows. Before you create even a single slide, take some time to sketch out on paper the basic structure of your presentation. Make sure you have an appropriate framework for your talk and a logical reason for any information you wish to present. So, stop what you're doing, turn away from the computer, and ask yourself three things:

1. What is the **objective** of my talk? (to highlight new data, give an overview of my research, get input from the audience?)
2. Which **main points** do I want to present?
3. What is the **key message** I want people to remember after my talk is over?

Make a list of the answers to these questions as the starting point for your presentation. Then sketch out your talk in draft form, using keywords and bullet points rather than complete sentences. After you've done this, review what you've written. Is your presentation logical and consistent? Are there extraneous pieces of information that can be left out? Are you trying to present too much information for the amount of time you've been allotted? As a general rule, 3–5 minutes should be spent on each slide, so calculate how many slides you should ideally have.

Identify Your Audience

Now that you've established your objectives and made a rough outline of your talk, the next thing to consider is your audience. How can you achieve your objectives given the knowledge level and interest of your audience? How well do they know the subject of your talk? If the audience is made up of people in your own research group, their knowledge level will be very high; if the audience includes other people in your department, the level might be somewhat lower. With a general audience their grasp of your particular subject will be even lower still.

Many graduate students make the mistake of assuming that they need to tell the audience everything they have done in the lab from the very beginning of their project. Not so. In fact, this is a common error, and you risk confusing people if you try to unload too much information on them at one time. Keep your talk short, simple, and to the point. It is not necessary to wow the audience with your productivity by telling them everything you've done so far. Your main message will just get lost in a mass of unnecessary details and digressions. So, the art of a good presentation is to share data as needed to support the storyline and leave out everything else. If it feels like you are 'killing your darlings', and you feel a burning need to add those slides to your talk, reflect on the storyline again. If that information/data is still not necessary to get your message across, save those slides for another presentation.

Once you've identified your audience, fill in the basic message of your talk with the appropriate supporting details. Don't be afraid to provide context or background information where necessary, or to explain the meaning of any acronyms—even if they seem obvious to you. This will be immensely appreciated by the people in your audience who do not know your subject as well as you do.

For your first couple of presentations, it isn't a bad idea to write out your talk to make sure you don't leave out any crucial information. Whatever you do, however, do not read from a script during the talk itself. This approach is guaranteed to put everyone asleep. It is also not a good idea to have your whole presentation written out as a prompt sheet. No matter how nervous you are, reading from a script is a disaster rather than a help. You'll end up speaking in a monotone, and your audience will be bored after just a few minutes and tune out everything you say.

Rehearse Your Presentation Aloud

You've structured your talk and made your slides. Now for the fun part: It's time to rehearse your presentation out loud. First to yourself (this will feel strange at first, but it is very effective for putting yourself at ease and for getting used to the sound of your voice in a quiet room). Then practice your talk in front of one or more fellow students or other trusted colleagues. Use these practice sessions to rehearse the pacing of your talk and to master the effective use of visual aids. Exaggerate your intonation a bit, as if you were telling a story to a small child or are an actor on the stage. Check with your peers to be sure you're not overdoing it. Most likely they won't have noticed your delivery as being exaggerated, but rather well-spoken and clear. Ask your colleagues for their comments and an honest assessment of your performance at the end of the presentation. Productive feedback from friends is useful for making improvements, and it's better to hear it from them rather than suffer the grumbles and complaints of strangers as they file out of the lecture hall.

Deliver the Presentation

On the day of the presentation, show up early and make sure you know how to use the equipment. If there's a microphone, find out how to turn it on and adjust the volume.

Okay. Now you're on. Greet the audience and tell them who you are (don't assume that everyone knows you, even in an informal setting). These introductory remarks have the additional purpose of getting the audience to settle down and direct their attention towards you. Clear presentations usually follow a standard formula:

1. In a sentence or two, tell the audience what you are going to tell them.
2. Tell them in detail.
3. At the end of the talk, tell them what you have told them.

The first part helps you to prepare the audience. By stating what you are going to talk about, you place your presentation into context. Next, give your talk as you practiced it, using your visual aids to support your words. Finally, sum up what you have told them, keeping your key message in mind and what it is you want them to remember after they've left the room.

Keep to your allotted time. If you've been given 20 minutes for your talk, then talk for 20 minutes, no more. Fifteen minutes is even better so that you can allow some time at the end of your presentation for questions and/or discussion. For many people, the question-and-answer session is the most nerve-racking part of the presentation. After all, you have no control over the questions asked, so you can't really prepare the answers. Or can you? A good exercise is to try to anticipate the questions you may be asked and prepare the answers in advance.

When you're asked a question, it's always a good idea to repeat or rephrase it to make sure everyone has heard it properly. That will also give you time to formulate an answer. Then go ahead and answer the question based on the data you presented (and on what you know). The next level of preparation is to have a few slides ready as a backup, with each slide answering one of the questions you anticipated. You can also put the question into a larger context by drawing upon data and information outside your own work. If you don't know the answer (and you're not expected to know everything!), or don't have enough data to support a proper answer, then say so. It's better to be honest than try to bluff your way through. The audience will notice this and question your credibility. If appropriate, tell the questioner you'll get back to them with more information when you have it. Learning how to answer questions can have a big impact on job interviews for your post-PhD career.

How to Give a Great Presentation: Tips for a Perfect Delivery

Public speaking is an art. Some people are great at it, others less so. But certain skills that will greatly enhance your ability to give a good presentation can be learned. Everyone loves to listen to a great speaker, so aim to be the kind of speaker you would enjoy listening to. During your presentation, your voice, facial expressions, and body language are your most important attributes.

1. **Be conscious of how you use your voice.** *How* you say it is just as important as *what* you say. Speak clearly and project to the back of the room. Don't rush. Use a natural pace, but don't be conversational. A monotone delivery is boring and will put people to sleep, so it's important to vary the pitch and speed of your voice as you talk.

2. **Pause at key points.** Creating space in your delivery at key points allows the audience to absorb your words.

3. **Look at the audience throughout your talk.** You will establish a rapport by making eye contact with as many people as possible. Just be careful not to fix your gaze on one individual, as this can be a bit unnerving. At the same time, be aware of your facial expressions. If you look and sound bored, the audience will be bored. If you are animated and alert, the audience will be interested in what you have to say.

4. **Don't talk to the projection screen behind you.** Address your remarks to the audience. Pay attention to the audience's body language and non-verbal reactions to your remarks. Know when to stop and when to omit part of your presentation if you sense that you've gone on too long, and the audience is losing the ability to pay attention.

5. **Avoid these annoying habits:**
   - Blocking the screen with your body
   - Excessive gesturing with your hands or moving about too much, such as pacing up and down
   - Mumbling and/or turning your back to the audience
   - Reading from your slides, word for word

Visual Aids

Giving a talk on scientific data is complex, so you will require the use of visual aids: charts, graphs, and tables will most likely form the core of your presentation. Using visual aids effectively will be as important to the success of your talk as your delivery. There is nothing worse than sitting through a presentation and being forced to look at slides that are badly made, indecipherable, unreadable, or have so much information crammed onto them that they are impossible to understand. Visual aids can be of many different types:

- PowerPoint slides
- Video and film (or links to them in your PowerPoint slides)
- Flipchart or whiteboard
- Molecular models or other 3D aids.

Whatever you decide to use for visual adds, keep it simple. Avoid switching from the whiteboard to PowerPoint slides during the same talk. This is confusing and distracting. While it may be an obvious point to make, do be sure you know how to operate the equipment you'll need beforehand.

Slides should contain the minimum amount of information necessary to get your point across: no more than three lines of bullet-pointed text, one graph or one table (with entries large enough to be read from the back of the room). Try to limit the number of words per slide to a maximum of 10–12. Use a minimum 18 pt Times Roman or Arial font for easy readability. Ensure each slide presents a coherent message: title (top), data (centre), and conclusion (bottom), and builds on the same story.

It's best to avoid copying graphs or diagrams from published reports for your slides. Redraw them so that they will be easily readable.

Be careful with using colour on your slides. The most readable slides use a dark blue background with white or yellow text.

Be aware of the room lighting. If there is too much light near the screen, it will be difficult to see the detail on your slides. Don't make the room too dark, however, or you risk having your audience fall asleep!

Take a look at **Chap**. 15 for tips and advice on successful remote meetings and presentations.

Common Mistakes

With the spotlight on you, it's tempting to try to impress your audience with an avalanche of data and plenty of bells and whistles. *See how much work I've done!* Nothing could be worse than this approach. In fact, this is a common error, and you risk confusing people if you overwhelm them with too much information. Keep your talk short, simple, and to the point. It is not necessary to wow the audience by giving them a minute-by-minute account of your prowess in the lab. Your main message will just get lost in a tangled thicket of unnecessary details and digressions. Less is more. So keep it simple.

Finally, no matter how nervous you may feel, relax, and try to look as though you're having the time of your life.

Group Meetings

A group meeting offers a more informal forum for presenting your research to your lab or department. It's still a presentation though, no matter how informal, so keep in mind the pointers and tips from the previous section.

How to Handle Your Group Meetings with Style

If your research group is typical, the person who heads up the lab will organise either weekly, bimonthly, or monthly group meetings, usually depending on how big the lab is. When it's your turn to discuss your work during the meeting, don't assume everyone in the group knows the exact nature of the problem you're working on. When it's your turn to talk, put that week's or month's problem in context so that everyone in the meeting is on the same page. Be sure to give credit where credit is due. If a student or colleague has contributed to your data, acknowledge their work. Listen carefully to your colleagues when they are speaking. Show them you are interested by asking pertinent questions.

Even if you're having difficulties in your project, try not to focus on the negative. Ask your colleagues for advice and support in an open and non-defensive way. Don't be afraid to admit you don't know something. Everyone is there to learn. No one can know everything.

Chairing a Session

If you are asked to lead a group meeting, this is your chance to sharpen your organisational skills. Keep the following in mind and you will impress your supervisor and labmates with your professionalism:

Chairing a meeting requires preparation. Every minute of preparation and planning you do before the meeting is well spent. Try to anticipate what might happen during the meeting and plan for any unanticipated obstacles. If you can anticipate (and eliminate) surprises in advance, you can deal with the core issues of the meeting more efficiently.

1. *Start on time*. This is a courtesy to those who bothered to show up at the meeting on time and sets a tone from the start that you and your group are serious.

2. *Stick to the planned agenda.* Everyone should have a copy of the meeting agenda. An agenda keeps the meeting on track by ruling out unrelated discussions. Everyone should have the opportunity to place an item on the agenda. Allow group members to submit agenda items in advance of the meeting.

Take a look at **Chap.** 9 for tips on collaborating with different personality types.

3. *Make sure each person has a chance to participate.* If you're not careful, some people will dominate the discussion, while others will leave the meeting feeling that their opinions and ideas were not heard. In particular, you might need to encourage the more introverted participants to speak up.

4. *Stick to the time frame.* Respect everyone's busy schedule by ending the meeting on time. If a discussion is becoming long-winded or is unresolved, ask the group members if they would prefer to extend the allotted time or to continue the discussion during a follow-up meeting.

5. *Keep to the rules of conduct during the meeting.* The rules of conduct for the group should be agreed upon by everyone and they should be adhered to. As chair, you need to keep control of the discussion and disallow any discourteous or disrespectful behaviour.

6. *Chair the meeting properly.* Your job is to monitor the meeting and make sure the agenda is adequately discussed. Do not abuse your position as chair to dominate the discussion with your own ideas and opinions. In most cases, your role will be to summarise the discussion. It is your job to make sure that the agenda issues are discussed, and the necessary decisions are made.

A Final Observation

Learning how to give a good presentation and conduct a successful meeting are important skills that will be useful to you no matter which career path you eventually follow. Take advantage of the opportunities to you as a graduate student to give as many presentations and run as many meetings as possible. Don't be shy and volunteer if necessary, as the more presentations you give, the easier it will be to stand in front of a group of people and make a lasting impact with your words.

Saving an Old Master Painting: Peter Gives a Presentation to the Art History Department

Two years into his PhD research, Peter has been asked by his supervisor to give a presentation to the Art History department on the progress he has made so far with the Lorenzo Monaco painting, *The Coronation of the Virgin*. Peter has given small presentations before in front of just a few people, but never for the entire department. He spends a great deal of time preparing his slides, as the visuals will be important when discussing the painting. He creates full-colour slides of the whole painting as well as various close-ups. He also makes copies of other Lorenzo Monaco paintings and those of his contemporaries for comparison. In addition, he consults Isabel and Yousef for some detailed information on the chemistry and physics of the degradation of the pigments. He starts the presentation off well by giving a brief introduction to the painting, putting it into historical and cultural context, and then outlining the problem.

Peter is excited to show the department the knowledge he has learned about the chemistry of paintings, so for the rest of the presentation, he launches into a complex discussion of a highly specialised nature (with the help of cue cards) involving the chemistry of pigments and their degradation products. He is so enthusiastic about showing off his knowledge in this area, that he doesn't notice the puzzled looks or fidgeting from the audience (he hasn't even bothered to explain what the acronym SIMS means). Furthermore, he has so many slides that he goes over his allotted time by ten minutes. When the presentation is over, Peter is disappointed that there are no questions. The next day Peter asks his supervisor for some feedback on the talk. His supervisor tells him that the talk was interesting but that he tried to present too much information. Peter has made the common error of trying to talk about everything he knows about this painting, and the chemistry of paintings in general, all in one go. He also failed to gauge the level of understanding of his audience and ended up dazzling them with complex chemical information that they were unable to fully understand or appreciate.

Though Peter disliked being criticised, as most of us do, he learned to keep the knowledge level of his audience in mind when preparing his next talk and only present enough information to support his key message.

On the next occasion to give a talk, Peter reduced the number of slides considerably, rehearsed the presentation with some labmates not involved in his project, and incorporated their feedback. This second talk worked out much better: it was well received, and he was offered some interesting suggestions during the Q&A session. One of the suggestions really helped him make some unexpected progress. The time spent on the preparation of the talk was repaid many times over.

13

Searching the Scientific Literature

The history of science knows scores of instances where an investigator was in the possession of all the important facts for a new theory, but simply failed to ask the right questions.
—Ernst Mayr

Before you carry out your first experiment in the lab, you would be well advised to spend some time in the library—and online—doing a thorough literature search of your research topic. Perhaps you worked on a project in this same field as an undergraduate, or think you are familiar with the subject area because it is related to other work you have done. Even if you have some knowledge of the literature on your project, you shouldn't skip this step. The investment of time in the library will pay off many times over in the lab. You don't want to risk doing work that someone else has already done or going down the same worn path others have travelled before you. Science is not carried out in a vacuum. It is about steady forward progress over long periods of time and wise graduate students will take the time to read and benefit from the research findings of their predecessors.

In recent years, the brick-and-mortar library has morphed into an online version that can be accessed on your computer from anywhere in the world. These days, there are few reasons to visit the physical library if what you're after is solely information. However, libraries still exist. Not only are they often majestic works of architecture, they're also great places to work—either on your own or in a small group.

© The Author(s), under exclusive license to Springer Nature
Switzerland AG 2022
P. Gosling and B. Noordam, *Mastering Your PhD*,
https://doi.org/10.1007/978-3-031-11417-5_13

As you embark on your literature search, you may feel quickly overwhelmed by the pile of papers you accumulate, so keep in mind that is impossible to read all the research ever published in your area. Being selective about what you read is key to getting a thorough overview of a particular field, without drowning in too much information.

But whatever you do, and however you decide to go about it, *do not skip this step*. You will live to regret it.

Getting Started in the Library—Both the Physical Space and with Online Search Tools

Get comfortable with the layout of the library and with the online research tools available at your institute. Introduce yourself to the reference librarian(s), or an experienced researcher from your team, and explain that you want to carry out a literature search on your thesis topic. These individuals are great sources of information and are there to help you in your search. You'll be spending a lot of time in the library so take time to get to know all its services. What kinds of scientific literature exist, and which ones will be most important to you? Broadly speaking, scientific literature can be divided into two types of articles: peer-reviewed and popular.

Most of the articles published in scientific journals, both primary (original research) and secondary (review articles) have gone through a stringent process called peer review. Before an editor will accept a paper for publication, he or she will send it out for review to at least two experts in the field. The identity of the reviewers is always kept secret from the author so that any comments will be impartial.

It is the author's responsibility to correct any errors or discrepancies in interpretation before the paper can be accepted for publication. This process, while not infallible, ensures that most articles are as free from error as possible. Articles published in popular scientific magazines are not subject to peer review, and thus not always reliable sources of information.

Secondary literature is published in the form of review articles. As the name suggests, these articles are often very comprehensive in nature and review the scientific findings in a particular field over a specified period of time. Reviews do not present new and original data, they are compilations of other people's work, but very often written by a 'big-name' scientist in the field. Review articles can be a goldmine of information and will significantly help you with your literature search by cutting down on the amount of time it takes to seek out individual (primary) research articles on your own.

Making the Most of Your Online Literature Search

At this point, you've already identified the scope of your research project (see Chap. 1, Getting Started). So, the next step is to take advantage of the wealth of information on the internet and do an electronic search on your research topic. Use key words wisely, however, or this type of search can get quickly out of hand. Start by searching popular and comprehensive databases such as MedLine, PubMed, GeoRef, and ScienceDirect. Your university library should have a subscription to some of these, as well as other databases that are password protected. Download and keep a file of the articles that are the most pertinent to your research. As you get more involved in your search, you will start to a get a feel for the important scientists in the field. Keep a file of their names and research institutions. Now is a good time to start archiving your references so that they're easily available in one location. To simplify this process, reference management tools are your friend. Two open-source, and easy-to-use reference management tools are Zotero and Mendeley. Both tools allow you to search, download, organise, and cite key research articles for your project, as well as collaborate with others online.

Another excellent way to get a jumpstart on your literature search is to read recent review articles published on your topic. Think of this as a bit of a shortcut. Someone, somewhere, has done much of the work before you and compiled it all in a comprehensive review article that may contain up to 200 references.

Another tip is to make printouts of the first papers published in the field and keep them in a file close to your desk (or as digital versions easily accessible on your computer). These papers represent the seminal work in your area of inquiry. It is critical to know how the field started, which experiments were done, and who the principal players were. As you search databases, identify review articles and key publications, you will begin to create a chronological picture of your research topic. It's very important to have an understanding of the early stages of inquiry into your area of research. As you read the papers chronologically up to the present, you will develop an understanding of how knowledge in your subject area proceeded. Surely in fits and starts, as science tends to go, but as additional research is carried out, more pieces of the puzzle are filled in.

Perhaps you're now beginning to understand why this process is essential. You'll need to be familiar with all the work that has been done on your topic, not just as a tool for learning, but to avoid repeating work that others have carried out before. Imagine the graduate student who spends six months doing a series of experiments only to discover (belatedly) that someone else has already done them, ten years ago. This happens more often than you might think, so don't let it happen to you.

Making Good Use of the Science Citation Index

This important tool can help you pick through literally hundreds of research articles to find the most cited and, hopefully, the most important articles in your area. Once you've gathered a solid collection of articles, you will need to scan through them and summarise and record pertinent information. How best to organise all this information? Keep a dedicated file of pertinent articles to build your bibliography.

For each article, make note of the author, title, name of the journal, and year of publication. Jot down a few words about each of the following:

1. Statement of the problem

2. Hypothesis

3. Theories and assumptions

4. Research methods

5. Data collection tools/procedures

6. Research design

7. Methods

8. Interpretation of the data (i.e., did the data support or reject the hypothesis?)

9. Conclusions/suggested future research.

If this sounds daunting and like a massive amount of work, just remember that investing time now in a proper literature search will save you vast amounts of time later on when you start writing your own articles based on your work in the lab (see Chap. 16), not to mention your thesis. You won't want to spend days and weeks in the library hunting down papers or finding out (oops!) that you've duplicated experiments someone else has already done, just when you're getting ready to write everything up.

How Do You Know When Your Literature Search is Successful and Complete?

You'll know you've made a comprehensive literature search when you have performed the following tasks:

- Identified the most recent articles (i.e., the past 10 years, plus seminal articles) on your research topic.

- Skimmed each article and prepared a brief summary of each one.

- Assessed each article for the strengths and weaknesses of the experimental setup, methods, procedures, data collection, and analysis.

It's up to you to develop an organised method for storing and retrieving this information. Many researchers file each publication (as a paper copy or in a file on their computer) and then append a cover sheet that includes a summary of the paper and an assessment of its key points. In addition, record this information on a master list, which includes the title of the paper, the author(s), and the name of the journal, etc. of each article, using a standard referencing format. This will save you oceans of time when you start writing up your own articles and have to refer to these references.

Even though it's no longer necessary to spend time in the library to carry out a literature search, libraries are still great places to hang out in. They can offer a much-needed refuge from the lab, provide state-of-the-art facilities to browse the literature, and employ dedicated library staff who can inform you about the latest search tools and strategies. Don't feel you're wasting your time if you enjoy spending time in the library while others are in the lab. Remind yourself that you might just know something they don't. Surely there's a wise saying that goes something like this: 'a day in the library could save you months in the lab.' Happy reading!

Saving an Old Master Painting: How Journal Clubs Can Enrich Your Research Project

Yousef is easily distracted by what he believes are urgent matters whenever he forces himself to read scientific articles related to his research. He considers reading the literature as a duty rather than a source of inspiration. A friend from university suggested he start a journal club. While it will not replace all the time invested in reading articles and keeping notes on the key findings, it does add a social aspect to the process. When Yousef approaches Isabel and Peter about starting a journal club to discuss seminal articles in their respective fields, they're both enthusiastic, but suggest they make a few ground rules before launching the club. First, they agree to discuss only articles all three of them are interested in. Second, all three must make an effort to read the article beforehand, while the person presenting the paper will read and digest the key points and peruse some of the references for additional context. Third, they will discuss three questions on the article: what was the hypothesis and key findings of the research; which aspects of the research described in the article might provide inspiration for their own projects; how is the article organised (including supportive tables and figures), and what can they learn from it? The newly formed journal club agrees to meet in the library in one of the smaller rooms where conversations are permitted.

While they did not use the library as a source of information, the group enjoyed being in a different setting. After discussing a single article at length and making notes on the aspects of their discussion and how it pertained to each of their projects, the club members treated themselves to lunch in the adjacent cafeteria. Their biweekly journal club was soon an event they all looked forward to. As an added bonus, Yousef has enjoyed reading articles related to his project much more since they started meeting together, but later realised they were discussing mostly review articles, while he would also have liked some in-depth discussion on new developments in instrumentation. While Isabel and Peter were interested in the topic, it was not part of their daily work, and they had limited experience with instrument development. Yousef realised that this was something he'd have to learn outside the club.

With no peers to reflect his insights in his own team, he happily accepted an invitation from a former Post-doc from his lab to join a virtual journal club organised by a group of five instrumentation experts scattered around the globe. After the initial stumbling block of finding a timeslot convenient for the time zones involved (Asia, Europe, US East coast), this second journal club was complementary to the one he'd formed with Isabel and Peter. Reading and discussing the literature has become a highpoint in Yousef's life, and something he plans on doing long after he obtains his PhD.

14

Your First International Conference

Human beings, by changing the inner attitudes of their minds, can change the outer aspects of their lives.
—William James

For quite some now you've been working on your research project in relative isolation. Interactions with the wider academic community have been limited to reading articles in scientific journals, and perhaps hearing some anecdotes about other groups from your supervisor or more senior PhD students. Now you're going to meet some of these big-name scientists at your first international conference. No doubt your excitement about the trip is mixed with a bit of apprehension, as presenting your own results was a requirement of attending the conference. So, you're not just going to listen to the big-name scientists, you'll also be sharing your own work with them. Moreover, just organising the details of the trip is turning out to be more of a hassle than you expected. This chapter aims to help you sort out all the things you need to make your first international conference a pleasure, rather than a pain, so that your experience of the conference is what it should be: one of the perks of working in academia.

Making the Most of Your First Conference

Attending a scientific conference is a valuable investment in your career as a graduate student—and beyond. In addition to your personal contributions to

© The Author(s), under exclusive license to Springer Nature Switzerland AG 2022
P. Gosling and B. Noordam, *Mastering Your PhD*,
https://doi.org/10.1007/978-3-031-11417-5_14

travel and other expenses, you'll be investing quite a bit of time at the conference. Altogether, attending a conference will take about two weeks of your time: one week to prepare and travel to the conference, and another week in attending the meeting. Fortunately, your investment will be well spent, as what you gain from attending the conference will be worth all the time and effort it takes to prepare for it. At a minimum, attendance at the conference will bring you up to date on the latest research findings in your field, you'll start building your network in the academic world, and the feedback you receive onsite will provide additional momentum to your research. Finally, attending a conference shares some features of a short holiday—particularly if the conference is held in an interesting location. But in order to get the most out of this potentially enjoyable investment, you'll have to put some effort into the preparation. This chapter will guide you through the key aspects of making the most of your first international conference.

First Things First

Selecting an appropriate conference to attend, particularly one that makes the best use of your time and energy, is a crucial decision. Your supervisor may even have suggested that you choose to attend a particular meeting. However, regardless of what your supervisor might have suggested, you should also consider attending the one that is most interesting and/or useful for you. Selecting the right conference can be based on just a few criteria. If you want to go to a conference that was not suggested by your supervisor, a strong rationale might help persuade your supervisor that your preference is the better choice. Here, we discuss four criteria for deciding which conference might be best for your entrée into the world of international scientific meetings.

First, it's important to make sure that your research fits well within the scope of the conference. This is a must. If the work presented at the conference has little relevance or overlap with your own work, the chance is remote that you'll get any practical tips or inspiration out of the meeting. If your work is a good fit, you'll enjoy the meeting much more. Especially when you realise that others are also interested in the type of scientific questions that you're addressing, and you'll be energised and inspired by their response to your work.

Second, make sure you'll be able to present your work in some form, whether as an oral or poster presentation. The fact that you haven't completed your research project should not stop you from presenting a work-in-progress type of poster. By presenting your research at the meeting, you'll shift your role from spectator to that of an active participant in the field.

Third, a conference is more than just a series of scientific presentations and poster sessions. Interactions with others in the field and networking with conference attendees are equally important features. Workshops and small conferences with less than 100 participants are best suited for getting to know other researchers in your area. Once you've established a network and have a sense of what others are doing in your line of inquiry, you might want to consider attending a bigger conference. Finally, if possible, try to pick a conference that's held in a pleasant location. After all, enjoying a conference can provide additional inspiration to continue your research. In a pleasant setting, interactions with others tend to go more smoothly. Even the top scientists in your field will be more relaxed, and perhaps even willing to involve you in scientific discussions. It helps if the location is somewhat isolated. If you're all gathered at a mountain venue or on a small island, everyone will stick around the conference venue, whereas in a major city, numerous diversions offsite will offer tempting distractions for you and the other attendees.

You probably imagine that organising your trip to the conference will be straightforward. You're not wrong about this: it is indeed a simple matter to register for the conference and book a plane ticket, but the list of details to be taken care of can be rather long. There's a bit more involved than buying a ticket and showing up at the airport. Without some degree of organisation, preparing for the conference may become quite time consuming. Two tips will help you organise your trip: work with a checklist and pay attention to things that require some lead time. On your checklist, make note of all the preparations you'll need to do, such as: find the best way to travel to the meeting, buy the tickets, register for the meeting, send your abstract to the organisers, make your presentation (either oral or a poster), bring pre-prints of your work and an A4-sized handout of your poster presentation, define the goals for the meeting, pack your stuff, and so on. Some of these things you can do on your own at any time. Others will involve interactions with other individuals and require some lead time. For instance, prior to sending your abstract to the conference you should share it with your co-authors for their comments. They might not respond immediately, making you nervous and irritated as the deadline for submission approaches. Ensuring that things run smoothly in your absence should be another item on your preparation list. Leave a note on your desk that informs others when you'll be back in the lab and how you prefer to be reached (phone, text, e-mail, etc.) if necessary. Put your e-mail in 'out-of-office' mode if you'd rather take a break from e-mail while at the conference. In addition to these small courtesies, you should also think of any projects that will be ongoing while you're away. Give proper instructions to an undergraduate student or lab technician who can operate the equipment or keep experiments running while you're away. Order materials from the stock room so they'll be available and ready to use as soon as you're back in the lab.

There are so many potential things to do at a conference, and so many possible distractions, that we suggest you create a short list of goals you'd like to reach during the meeting. Such goals can be related to your own presentation (such as asking the audience for input on a particular experiment), or faithfully attending the presentations that are key for your research or building your network. For example, introduce yourself to a member of another research group and ask them how they prepare their samples. During the meeting you'll likely find that there are plenty of interesting things to do, none of which you'd anticipated. Nevertheless, check your list of goals every now and then to make sure you're staying on target, and then consider how to execute your original plan. How you handle the often-overwhelming conference programme is an art in itself. A large conference will have plenary lectures, keynote speakers, parallel sessions, multiple poster sessions, and industry booths, etc.

Prior to the conference, you'll have likely browsed the website with a summary of the programme. Once you've arrived at the conference site, you may be feeling enthusiastic about attending the first session on Monday morning. Chances are, you'll enjoy quite a few of the lectures, while others may be less interesting and of little relevance. Perhaps the Monday evening poster session appears only marginally relevant to your research. But you wander through the poster presentations, anyway, since you're determined to get as much out of the conference as possible. But let's suppose that by Tuesday afternoon you've faithfully attended every presentation and are beginning to feel a bit overwhelmed. Due to a touch of brain overload, you may even be having a hard time paying attention to the lectures, no matter how interesting they are. Perhaps you're feeling tired at this point and have already absorbed so much new information that you lack the energy to chat with others during the breaks.

By the second half of the week, the amount of information you're able to absorb has likely shrunk considerably. Too bad, then, that you spent some down time in your hotel room and missed the very interesting and relevant talks on Thursday afternoon. You realise, perhaps too late, that a week at a conference can be a very long time indeed. Our advice is simple. Follow the tactics of a marathon runner and do not start out at full speed. Pace yourself and carefully plan your attendance at presentations. You are not being graded on how many lectures you attend. It may seem obvious but save your energy for more important things by choosing not to attend presentations that have little relevance to your own work. Our advice for getting the most out of the full week of presentations and interactions is to skip a good fraction of the programme, particularly the sessions that have little relevance to your own work. Use that time to relax, talk to others, and to digest all the information you have taken in so far.

As stated previously, we believe that building a professional network is one of the major reasons for going to a conference. That's why the coffee and meal breaks can be as important as the presentations themselves. Presentations usually highlight what has worked in a particular line of research. In other words, at a conference people tend to showcase their successes. It is only during the breaks that you will find out about any attempts that did not work out. In retrospect, newly minted PhDs often realise that most of their efforts during their PhD research did not contribute to the work reported in their thesis (see also the section in Chap. 3 on the 80/20 rule). During informal discussions you can learn from others about what not to do, as well as strengthen your sense of community and to realise that you're not working on a problem in isolation. The kind of information you discover can be mundane, or it can be as important as learning that the type of sample preparation you've been attempting is worthless.

By sharing failures and setbacks with others, you will also receive useful feedback on your research project. A second reason to work on your network is that wild ideas about new research directions and collaborations on future research often get their start during conference breaks. In addition, when you're finished with your PhD you might want to continue doing research in another academic group as a Post-doc. Meeting people from research groups in different countries will allow you to make a more informed decision about where to go next. You might also collect a few interesting stories about former PhD students who have left academia. Finally, informal chitchat can be a welcome interruption from staring for hours at all those PowerPoint presentations. So, we'll remind you again: don't forget to enjoy the meeting. Having a cappuccino on the steps of the conference building can be more productive and enjoyable than fading away in boredom at yet another incomprehensible lecture.

Making Your Presence Count

Most likely you will be presenting a poster rather than giving an oral presentation at your first conference. It might seem to you that poster presentations are a minor aspect of the proceedings and not very important to focus on. During these sessions, however, there is a great deal of personal interaction, and they can be extremely rewarding. In oral presentations given in front of a large audience, there is often little response to the work presented, aside from a question or two from the audience. The interactions and discussions that occur during poster sessions might yield a couple of valuable suggestions for moving forward in your research, either at present or in the future.

At smaller conferences, the posters are often displayed throughout the meeting in the same area where coffee is served. In this case your poster will get quite a lot of exposure. Naturally, you cannot stand by your poster at all times, so be sure your poster is self-explanatory, with a clear introduction, methods and results sections, and clearly stated conclusions. Make up a stack of A4-sized sheets of your poster and place them in a folder that is tacked to the bulletin board to which your poster is affixed so that people can take a copy with them. Also, it's a good idea to put your photograph somewhere on your poster. This helps people to find you later on if they want to discuss your research with you.

Choosing which presentations to attend from the myriad on offer requires some strategy. Before even entering the lecture hall, decide whether this particular presentation will be of interest to you. If you've chosen unwisely or find yourself listening to a presentation that has little relevance to your research, do not despair or leave the lecture in frustration. But trying to follow the lecture at this point will just be a drain on your energy, so the best tactic is to ignore the presentation altogether and let your mind wander. Some of the bigger conferences have parallel sessions, and if you will not be too obtrusive, you might try to slip out a side door so that you can attend another presentation.

If, however, you find yourself listening to an interesting presentation that has great relevance to your work, make an effort to focus your concentration on the message and key points. Two things can help. First of all, make notes. By writing down the key points, it will become clear what you understand and what you do not. Second, prepare a question. Possibly the session chair will not allow questions due to time constraints, or perhaps you won't dare to raise your hand. Nevertheless, you should prepare a question anyway. It will make you a better listener. Later on, during the break, there might be an opportunity to discuss the question with the speaker (most speakers love it when people approach them after their presentation), or, if that's not possible, with someone from the same research group.

If you've put a lot of effort into reading the programme, you might be tempted to zigzag through the offerings so that you can attend all your favourite talks. We strongly advise you to hop as little as possible from one parallel session to the other. The session chairs seldom stick to the schedule, and you might feel frustrated if you enter the lecture hall after, rather than before, the start of your favourite talk.

We hope you've been having a good time at your first international conference. But no matter how enjoyable you find the conference to be, remember that it is distinctly different from a vacation. Do socialise at the bar, but do not become so caught up in the conviviality of the moment that you are tempted to skip the first interesting lecture in the morning. Organisers know that people tend to straggle in late, so they try to schedule the most interesting talks for the first session of the day.

Post-conference Reality Check

Finally, the conference is over and you're back on home ground. You'll likely need to take care of an overwhelming number of things now that you're back at your desk and in the lab. Your inbox is overflowing with e-mails; you've got to teach a class later in the week; and you'd like to socialise with friends you haven't seen for a while. In short, within a day or two you might have forgotten all about the conference—even all those great new ideas you picked up while in attendance. Before the conference becomes a distant memory, take the time to go back through the conference programme and look at your notes. Find a convenient place to keep the business cards and phone numbers you collected. Offer to run the next group meeting or give a departmental talk in which you summarise the highlights of the presentations you attended. Those unable to attend will highly appreciate this gesture on your part, and it will provide an opportunity to discuss with others the research findings that sparked your interest.

Saving an Old Master Painting: How Peter Submitted His Conference Proceeding

Peter is planning to submit an abstract to a large international conference. Based on the submitted abstracts, the conference organisers will decide whom to invite to participate in the conference or give a presentation at the meeting. As planned, Peter wrote a first draft of the abstract and gave it to his supervisor. Peter's supervisor agreed to look at it as soon as possible. Then an emergency came up and the supervisor had to put his abstract aside. Unfortunately, Peter's supervisor is not very well organised (most likely the cause of the emergency in the first place) and has many other things to do. So, after fixing the emergency, he forgot about Peter's abstract, which is now lost on his desk, or at the bottom of a pile of other things to do. At the end of the month, Peter has not only not submitted the abstract, but he has also lost quite some time in chasing after his supervisor. Even worse, because Peter is irritated by all this wasted energy, he is less effective in carrying out other projects. The irritating obstacle of the conference abstract has to be overcome. It would be a shame if Peter missed out on attending the meeting just because his abstract had not arrived in time. Now that he has identified the hurdles, a solution might be to confront his supervisor with the deadline and ask his permission to send it out without his review, in the event he has no time to look at it.

Such a direct approach has two drawbacks, however. First, most supervisors do not like to be told what to do, so some sort of indirect gesture needs to be made. Second, without input from others, the abstract might not be good enough to be accepted by the conference organisers. Someone has to read the abstract to help him improve it. Therefore, Peter asked Isabel and Yousef to read the abstract as if they were the supervisor and to come up with suggestions, in the same way they trained themselves how to review articles in their journal club (see Chap. 13). Peter approached his supervisor and told him that he understood that he had little time to look at his abstract, and that the abstract could not be sent out without his permission. Peter mentioned his solution and repeated the useful suggestions made by Isabel and Yousef. Now his supervisor looked briefly at the abstract (which he miraculously picked out of a seemingly disorganised pile) and agreed to submit the paper. Peter has not only managed to finish the abstract, but he also identified a solution to make it happen. He feels more in control of the situation, and his pro-active and constructive behaviour has made it much more likely that his abstract will get submitted on time and be approved. In the end, Peter's supervisor was pleased with the practical approach Peter took to increase the chance of getting the group's paper accepted at the conference.

15

Remote Collaboration

A system of telephony without wires seems one of the interesting possibilities, and the distance on the earth through which it is possible to speak is theoretically limited only by the curvation of the earth.
—John J. Carty, 1891

Long before the start of the Covid-19 pandemic, which brought an abrupt halt to onsite and face-to-face meetings that lasted for two-plus years, there were video, chat, and other online platforms for collaborating on a research project. However, before the pandemic threw a spanner into the works, team meetings and conferences were traditionally organised as live events that took place at your institute or a conference centre, often with participants flying in from across the globe. In the early months of 2020, however, with many countries in lockdown to prevent the spread of Sars-CoV-2, it became clear we had to find new ways of working together—while remaining apart. Online and remote collaboration tools rapidly improved; we learned how to cope with the frustrations, limitations, and benefits of remote work; and now, more than two years later, remote meetings and online networking opportunities have become essential tools for collaborating with colleagues, whether they're located in an office down the corridor or on the other side of the globe.

In this chapter, we discuss how you can benefit from the rapid adoption of remote collaboration tools and suggest ways of dealing with both the up- and downsides of meetings that are no longer face-to-face. Unlike gathering in person, a virtual meeting doesn't require you to travel from your home to a different location—with the benefit of saving both time and money. It's

© The Author(s), under exclusive license to Springer Nature Switzerland AG 2022
P. Gosling and B. Noordam, *Mastering Your PhD*,
https://doi.org/10.1007/978-3-031-11417-5_15

just you, alone in a room, sitting in front of your screen. Nevertheless, the dynamics of online meetings change with the number of participants you'll be interacting with. Therefore, to make these meetings as effective and engaging as possible (as opposed to stultifying and draining) in our brave new world, adapting the way you interact in a remote setting is a must. We'll cover the aspects of online meetings in order of the number of attendees, from low to high, and finish off with advice on how to handle one of the most challenging forms of collaboration: the hybrid meeting.

Meeting of One

Although vastly different from face-to-face meetings, virtual meetings share several social ground rules. In the pre-Covid era, when attendees gathered in the same room, the meeting start-time—in some cultures, but not all— typically included a five-minute grace period. No one was likely to complain if you arrived a few minutes late or a few minutes early. If you happened to arrive early, you could enjoy the extra few minutes by getting a coffee or checking your email. If you were late, you might excuse yourself briefly to the meeting chair before slipping into your seat. The same 5-minutes grace period typically holds for one-to-one virtual meetings, since there's only one person who might be mildly annoyed at your tardiness. In a one-to-one virtual meeting, the other person may only be a click away, but try to keep your annoyance in check—especially if you're the type who is habitually punctual—if that click doesn't occur strictly on the hour. Disruptions happen, particularly when people are joining the meeting from home and sharing their household with others. However, for larger groups, grace periods for tardiness tend to be less lenient. A colloquium will usually start only a few minutes late, if at all, and virtual conferences with many attendees (and a high probability of technical problems), typically start their presentations on time—primarily to avoid chaos—and keep strictly to the agenda.

Although sitting behind a screen may have a distancing effect compared to meeting face-to-face, don't forget to extend a cordial greeting before you get down to business: 'Thanks for inviting me to this meeting/conference/presentation. I'm looking forward to it.' Since many people plan meetings back-to-back, with little or no time in between, respect your allocated time slot, whether you're the organiser or a participant. If you think you'll need some extra time for your presentation or input, give the meeting chair plenty of warning, just as you would in a meeting taking place in a conference room.

A potential drawback of virtual meetings is the loss of non-verbal cues. Thus, it's important to ensure during a virtual meeting that you respect each other's personality traits and temperaments (see Chap. 9: How to get along with your labmates, et al.). If you're chairing the meeting, be sure to give the introverts time to speak, and to reign in the extroverts, as needed, if they're monopolising the discussion.

All these different factors tend to go more smoothly if you've already met in person the people you'll be interacting with in a virtual setting. If the remote meeting is the first time you'll meet the other attendees, spend a bit of extra time preparing for and respecting the usual social ground rules. For instance, if you meet an expert in your field from whom you'd like to seek input or feedback on your research results, the success of the meeting is in your interest, so you'll be the one doing the prep work beforehand. Prior to the meeting, send some data or a short presentation by e-mail—but avoid sending an overwhelming or incomprehensive data dump. Start the meeting by thanking them for their time and close the meeting—on time— by thanking them again and asking for feedback. A final question such as 'I know our time is almost up, so I'd like to ask if there's anything particular in my analysis of the data that you're missing?' may trigger some thoughts. Although such efforts might feel a bit forced or unnatural in the remote setting, when all you have to work with are pixels on a screen, being clear— and confident—about what you want will help you get the information you need.

Remote Work Sessions

If you're writing a journal article with a former PhD student from your lab who's now doing a Post-doc elsewhere, you'll likely need to meet regularly to discuss progress and resolve any issues. Since you already know each other well and have had plenty of face-to-face interactions in the past, these video chats or remote meetings can have a very informal character. Such a collaborative, real-time approach will allow you to tackle difficult problems or brainstorm how best to solve a key issue. As a point of contrast, have you ever had a fruitful brainstorming session by e-mail? Probably not. To bridge the gap between face-to-face and remote collaboration, make a point of keeping your remote work sessions informal. Schedule short breaks to grab a cold drink or a coffee and take a few minutes to stretch your legs. One of the upsides of remote communication—in addition to saving money and travel time—is the ability to mute your microphone and switch off your camera, so you can take a couple of minutes for an informal break.

Team Meetings

Virtual team meetings with a medium-sized group are most beneficial when all participants make an effort to have a productive meeting. In a small gathering (e.g., 4–5 participants), you truly need to participate to make the meeting work. Sitting back and listening is not really an option in this case. Meetings with larger groups typically rely on established relationships for their success. Large or small, everyone needs to participate for the meeting to be a success. Large groups benefit from having an active chairperson who keeps the meeting on track and to interject if one person is dominating the discussion. In a smaller meeting that lacks a chairperson, perhaps one person might chime in to remind the team about an earlier question that wasn't addressed, and another might suggest it's time for a short break. Someone else may notice the discussion is drifting away from the original focus of the meeting and suggest they get back on track or continue exploring the other interesting items on the agenda.

Just like face-to-face meetings, remote meetings require some preparation. Make sure everyone has received the agenda and any supporting material ahead of time. If you're chairing the meeting, make sure everyone has a chance to participate and express their opinion. Particularly with larger groups, respect each other's schedule by starting and finishing the meeting on time. If you need a few extra minutes at the end, check first that exceeding the time limit is okay with all attendees.

Department Meetings

Larger meetings with more than ten participants, typically have a different setup. Examples include departmental colloquia with a professor addressing the whole group. In such meetings, most participants are receiving information, while only a small fraction actively contributes: the speaker, the chair, and sometimes a panel. These larger meetings, with passive listening from the attendees, can be a bit tedious, even more so than the in-person version of the—often mandatory—department meetings, in which administrative minutiae are typically discussed. But hang in there and try to stay actively engaged. These types of meetings are part and parcel of being a member of an academic institution, and if you drift off, you might miss something important.

If you're the speaker at the virtual departmental colloquium, you may find you get little feedback from the audience. If you happen to be the presenter (e.g., for the annual update of your research project to the department), you can actively seek feedback by pausing and asking for input, relating a funny anecdote about something that happened in the lab, or by quizzing the audience about a particular event. If the energy of the group appears to be flagging, the meeting chair can call for a coffee break in between agenda items to restore the energy of the audience.

Most likely, as a PhD student, you'll be a member of the audience. Your brain may be fighting to concentrate on the presentation. As time passes, however, your thoughts may wander to more interesting topics, like what you'll eat for lunch, or if that important e-mail has shown up in your inbox yet. Since you're mostly staring at a screen, remote meetings can be tiring, particularly if you're obliged to passively listen for long periods of time. We suggest either focusing on the presentation if you find it interesting, or completely disconnecting to give your brain a rest. Everything in between is exhausting and of little use for staying engaged.

© Herman Roozen/ComicHouse.nl

Virtual Conferences

Do not underestimate the challenges involved in participating in a virtual conference. Yes, you've been spared the need to travel. Yes, you can wear your pyjamas or workout clothes and hop in and out of the meetings at will. There is no need to rush down the maze of hallways in a big convention centre to get to the session on time, only to find yourself standing in an overcrowded lecture hall for an hour or more. On the other hand, you will not have serendipitous encounters with former labmates or get into a casual conversation with an expert in your field while waiting in line at the coffee kiosk.

The biggest pitfall of attending a virtual conference is the tendency to click eagerly from one session to the next, without giving yourself a break or to move around a bit. Thus, on the first day of the 4- or 5-day conference, you're already exhausted. A bit overwhelmed by all the information, some presentations may inspire you to take the next step in your research project; others seem interesting but are also difficult to take in all at once. One way of sitting through such conferences is to attend the screening with some of your labmates. Meet at someone's home and have dinner together— just as you would have done at a traditional conference. Skip a lecture that doesn't interest you and go outside for a walk in the fresh air. In this setting, surrounded by others from your lab or department, you'll not only be absorbing new information and learning interesting things, but doing so as a team. Later on, you can digest everything you've learned into manageable action items for your research project.

Hybrid Meetings

The most difficult of all meetings are arguably the so-called hybrid meetings: some participants are in the physical meeting room, while others dial-in. It's a difficult setup for creating an equally collaborative meeting for the whole team. The biggest pitfall is that the participants in the room will likely have a lively and engaging meeting, whereas those dialling in might feel more like an observer than a participant. It's important that the meeting chair balances the contributions between those in the room and the virtual participants. You might want to take turns in learning how to achieve this balancing act, so you get to experience both scenarios and learn how this works. Chairing a meeting—whether in person or remotely—is one of the important skills you should acquire during your PhD training.

To balance the contributions and interactions of the individuals in the room and those who join remotely, you may ask all participants to have their laptops and cameras on, so everyone's facial expressions can be seen. A wide-angle camera in the meeting room will show some of the group dynamics, but when used without close-up individual cameras, more is lost than gained. Those in the room can either look at their screens to see the remote participants or look away from their screens to interact with others in the meeting room. Hybrid meetings, whether we like them or not, have become an accepted feature of collaborating or doing business. Even if a global pandemic is no longer a factor in keeping everybody at home, remote work and virtual meetings are here to stay.

Saving an Old Master Painting: Isabel Learns How to Fix an Old X-ray Camera with Over-the-Shoulder Support

While reviewing an X-ray image, Isabel and Yousef notice that the left bottom corner of the picture shows ambiguous results. The image is blurry, apparently caused by an error in the camera. Isabel had taken the picture a year ago, and she regrets now that she'd only analysed the raw data. They agree that they will have to retake the picture. In the meantime, others have used the equipment in the lab, and Isabel has to scramble to set everything up. The X-ray camera is quite old, and she discovers that it's no longer functioning. To make matters worse, the instruction manual is missing. After consulting the group's Wiki, however, she learns that the filament for the X-ray tube, which produces the X-ray radiation, most likely needs to be replaced. Now what? Where is she supposed to find a spare filament and learn how to replace the broken one? Isabel initially spends valuable time wishing her group were better funded so they could just buy a new camera, but later realises her irritation is a waste of energy. Fortunately, she's able to contact Katrien, who'd written the original Wiki instructions, on LinkedIn. Katrien advises Isabel not to replace the filament by herself, as it wouldn't be safe and should be carried out by the manufacturer. Moreover, Katrien suggests that Isabel perform a simple test first: check the safety light connected to the X-ray setup. If it's defective, the X-ray tube won't switch on. In order to do the test, she'll first need to replace the safety light. Isabel is lucky to find a spare one stored in the cabinet. The instruction manual is missing, however, so replacing the light will not be straightforward. With the help of FaceTime on her smartphone, Isabel is able to walk through the setup with Katrien, who explains how to replace the light, step-by-step. In the end, Isabel is able to replace the light and get the X-ray setup working again. Remote learning to the rescue! Without this 'over-the-shoulder' support it would have been difficult—if not impossible—for Isabel to repair the old setup and to take a key X-ray image of the painting she needs for her project.

16

From Data to Manuscript: Writing Scientific Papers That Shine

The scientist is not a person who gives the right answers, (s)he's the one who asks the right questions.
—Claude Lévi-Strauss

At long last, you've completed a series of experiments and collected enough data to write up your findings as a scientific article for submission to a peer-reviewed journal in your field. At this point in your career, you've read dozens of scientific papers and are familiar with the format. Following the standard format and tailoring it to your own work is relatively easy if you keep the following points in mind as you write:

A scientific article is a written document of your work in the lab or in the field. Keep in mind that its purpose is to communicate your research to the scientific community and to provide researchers in your field with specific types of information:

- Which questions did you ask?
- Which experiments did you perform to answer these questions?
- What types of data did you collect and how was it collected?
- Which conclusions did you draw from your data and what suggestions have you made for further research?

Before we get into the particulars, however, a few general points should be kept in mind while you're writing. Research demands accuracy and precision. Scientific writing should reflect this in the form of clarity. Unfortunately,

© The Author(s), under exclusive license to Springer Nature
Switzerland AG 2022
P. Gosling and B. Noordam, *Mastering Your PhD*,
https://doi.org/10.1007/978-3-031-11417-5_16

if you glance at almost any scientific journal you will discover that clarity and concise writing are very often lacking. Many of the complaints by non-scientists of obscurity and elitism within the scientific community partly stem from the fact that many scientists are incapable of expressing their hypotheses and conclusions clearly and simply. Don't allow yourself to fall into this trap. Part of being a good scientist is not just designing good experiments but being able to present your work and write it up in clear and simple language. Obscure language will not make you sound more intelligent; it will only confuse others. As a result, your work will have much less impact on your intended audience.

A well-written scientific article will provide answers to the questions listed above. The standard format used in nearly all peer-reviewed journals will help you organise your material into a logical order. Take a look at any paper from a respected journal in your field and you will see that it's been divided into the following sections:

- Abstract
- Introduction
- Materials and Methods
- Results
- Discussion
- Conclusion

Title

A good title is an art in itself. Give your article a strong title for maximum impact. Try to create a 'dynamic' title rather than a 'static' one. A dynamic title contains the key result from the study: *'Cyclophosphamide inhibits tumorigenesis by blocking the phosphorylation of protein zeta'*. Whereas a static title is merely descriptive: *'The role of cyclophosphamide in tumorigenesis'*. See the difference?

Abstract

The abstract is a one-paragraph summary (typically 200–300 words) of the research described in the article. It should be a self-contained summary that is complete enough for the reader to understand the purpose of the research and the results without having to read the entire article. The abstract should contain the following elements:

1. The central question (purpose) of the study
2. A brief statement on how the study was carried out (methods)
3. A brief statement of the results found
4. A brief statement of the conclusions.

Note that many computer search algorithms make use of the information in the abstract. Make sure, therefore, that you have the relevant key words in your abstract so that your article will be easy to find by internet search engines.

Introduction

For many people, this is the most difficult part of the paper to write. Deceptively simple, the introduction must contain a great deal of information in a short amount of space. This means you'll need to write crisp and concise sentences to put your work into context. It's important to include enough background so that a reader not familiar with the field can understand the relevance of your work.

The purpose of the Introduction is to explain to the reader why you decided to conduct your research. So, this is the place in which you answer the following: which questions were you attempting to answer? State any information about previous related research or existing knowledge in the field. How did the information that already exists help you in planning your own experiments? In other words, the reader of your article wants to know: why did you, the researcher, do this work? Be sure to clearly state your hypothesis and objectives. Read the introductions of several well-written papers to get an idea of the content and style. Some journals allow you to write the main conclusion at the end of the introduction. Make use of this opportunity when you can as it will prepare the reader for what's to come in the main body of your article.

Materials and Methods

In the Materials and Methods section, you will provide a clear description of exactly what you did and how you did it. This section is extremely important, and details count. What was your experimental setup? Which type and brand of equipment did you use to collect your data. How and when was the equipment calibrated. Which chemicals did you use (including the company you ordered them from, and even the batch number, are important pieces of information). Keep in mind as you write up this section, that you will need to provide enough information so that other researchers can understand exactly what you did and will be able to duplicate your work. Again, study several well-written articles from respected journals to get a sense of what to include in this section, and the style that other authors adopted. Also note that it's common practice to describe the methods using the passive voice: 'The pigment sample was heated to 50 °C' rather than 'We heated the pigment sample to 50 °C.'

Results

The Results follow logically from the Materials and Methods section, as it's the section where you present the data you've collected. Not data in its raw form, but analysed data, which is usually displayed best in graphic or tabular form for ease in presentation and interpretation. Particularly if your collected data resulted in a lot of numbers, you will need to determine the best way to present it. A combination of tables and graphs is usually a good choice so that the reader can see both the numbers and a graphical presentation of the relationship between two variables. The Results section must closely match the techniques and procedures you described in your Materials and Methods section. For example, if you present temperature data in the Results section, then the Materials and Methods section should state when and how you measured the temperatures you obtained.

A Note on Tables and Figures

Tables and figures are used to convey data in a more efficient way. Both tables and figures must be able to stand alone and should be accompanied by an explanatory caption that allows the reader to understood them without having to read the body of the paper. Avoid repeating in the body of the manuscript information that' s in the captions of tables or figures. Do refer to the data in the figures and tables, however, when appropriate.

Tables

Do not repeat information in a table that you have already depicted in a graph or histogram; include a table only if it presents new information, or the exact value of the certain measurements is relevant to your results. It is much easier to compare numbers by reading down a column rather than across a row. Therefore, list the sets of data you want your reader to compare in vertical form. Provide each table with a number (Table 1, Table 2, etc.) and a title. The numbered title is placed above the table.

Figures

Figures can be graphs, histograms, spectral traces, etc. Provide each figure with a number (Fig. 1, Fig. 2, etc.) and a caption that explains what the figure illustrates. The numbered caption is placed below the figure.

Graphs and Histograms

Both graphs and histograms can be used to compare two variables. However, graphs show continuous change, whereas histograms show discrete variables only. Decide which is the best way to represent your data. You can compare groups of data by plotting two or even three lines on one graph, but avoid cluttered graphs that are hard to read, and do not make the (common) mistake of plotting unrelated trends on the same graph.

When creating graphs or histograms, plot the independent variable on the horizontal (x) axis and the dependent variable on the vertical (y) axis. Be sure to label both axes, including the appropriate units of measurement.

Tips for Making Great Graphs

1. At first glance, your graphs might look good enough, but it's important to make sure that they aren't confusing, and that they don't have complicated axes or extrapolations that claim more than they should. Here, we offer a few tips for making graphs that are both accurate and 'honest'. Whenever possible, begin your axes at zero and use appropriate scaling. Sometimes, however, a valid trend will disappear on a graph with a zero axis, and all

the data points will bunch together at that top. In such a case, inform your readers that the graph's axis is not zero, either by stating this in the text, or with a clear break in the axis.

2. If a data point represents the mean from a number of observations or experiments, indicate the variability by a vertical line through each point and indicate whether this is standard error of the mean or standard deviation. In addition, specify the number of observations or sample sizes.

3. When comparing two graphs, make sure to draw them both to the same scale for ease of comparison. If possible, place them side by side in the article.

4. Be aware of the limitations of your data. Extrapolating a line or curve beyond the points shown on the graph may mislead the reader.

5. Pay particular care with line graphs. A false impression of your data may be given if successive points are connected by lines. It may be better to present the data as a histogram, or to leave the points on the graph unconnected by a line.

Drawings and Photographs

Pictorial forms, including drawings, photographs, and other images, are used to illustrate organisms, experimental apparatus, models, cellular and sub-cellular structure, and results from techniques such as gel electrophoresis or electron-microscopy. Preparing such figures can be time consuming as well as costly, so be sure that each drawing or photograph adds enough value to your article to justify its preparation and publication. On the other hand, a good illustration that shows the key result in your article can be of great value. Moreover, if you use an eye-catching illustration on the first page to highlight your results, it will act as a focal point for your paper.

Discussion

In the Discussion section of your article, you'll present your interpretation of the data presented in the Results section. You're allowed a little leeway here, but don't get too carried away with assumptions and wishful thinking. Be prepared to back up your analysis with solid evidence as provided in the Results section. Take care that you don't include in your analysis any data that you neglected to mention in the Results. Journal editors and reviewers are trained to look for these types of discrepancies. In a nutshell, the purpose of the Discussion section is to explain the meaning of the results. For example: did temperature effect the rate of decomposition of a particular pigment? Don't make the common mistake of confusing the Results section with the Discussion section. The Results section contains only the data you obtained from measurable parameters. The Discussion section explains the relationships observed in these data. Thus, any patterns you discovered, based on your data, are described in the Discussion section.

In addition, the Discussion section provides space for you to answer the questions that were posed in the Introduction (and that may have arisen in readers' minds as they read your paper). In other words, did you discover what you thought you would (i.e., did your experiments prove or disprove your hypothesis?) Were the results different from what you expected? What have you learned from your analysis? How does your work relate to other work in the field? Does it confirm or refute existing information?

The Discussion section is also the place for suggesting ideas for future research. You may have answered some of the questions you started out asking, but most certainly the work you carried out has led to new lines of inquiry. Pose those new questions here. In addition to your own work, they may provide possible new leads for other researchers in the field.

Literature Citations

A reference list of cited literature is the last section of the paper. Be sure to follow the referencing style as described in the Instructions to Authors for the journal you plan to submit your paper to. Within the body of the text, you must cite another researcher's published work whenever you refer to his or her results, conclusions, or methods in your paper. In general, a reference to a particular study in the text includes only the first author's surname (plus 'et al.' if relevant) and date of publication. There are three common ways of doing this:

1. Both the name and date appear inside parentheses, if the name is not actually part of your sentence: (Smith et al. 2021).
2. The first author and year (in parentheses) is typically placed at the end of the sentence or clause containing the citation. Any necessary punctuation comes after the citation: It has been shown that UV light can severely degrade cobalt-containing pigments (Meyer et al. 1968).

3. Another way to cite a study is to use the last name of the researcher (plus 'et al.' when appropriate) as the subject or object of the sentence or clause and follow it immediately with the date of the study in parentheses:

- Holloway (1993) found that cobalt-containing pigments degrade in UV light.
- Ever since Carvello et al. (2014) characterised the degradation products of cobalt-containing pigments, due to exposure to UV light, numerous researchers have investigated the UV-degradation of a range of pigments.
- These data support the conclusions by Jacobsen et al. (2017) as described in their study on UV and visible light degradation of oil-based pigments.

If you wish to emphasise the date of the cited study, you can omit the parentheses: As early as 1968, Meyer and colleagues showed that UV light can severely degrade cobalt-containing pigments [1].

This strategy is often effective for presenting a historical perspective of the problem, which can be useful for describing background information in the Introduction section.

Note, however, that it is never correct to separate the date of publication from the author's name, so avoid statements such as the following: Holloway found that UV light degrades cobalt-containing pigments (1977).

If you wish to cite more than one study per reference citation, i.e., if more than one author has reached the same conclusion or worked on the same problem independently, you may list them together in the same parentheses and separate their names by semicolons: UV light has been shown to degrade cobalt-containing pigments (Meyer 1968; Holloway 1993). By convention, these citations are usually listed in chronological order.

Revising the First Draft

Once you have written the entire article in the format described above, it's time to take a well-deserved break. Congratulations on a job well done! But you're not finished yet, you've only written the first draft. Print out your article or close the file and avoid looking at it for a few days to get some much-needed distance from the process. In the next stage, you'll be switching hats: from that of writer to editor. After a few days, when you're ready to return to your article, choose a time and place where you won't be distracted and read it all the way through with a cold and critical eye (just like a reviewer and eventual reader will do). Don't be lazy about this step as you'll just delay publication. Any sloppiness on your part will be spotted by your peer-reviewers and the paper will be sent back to you for corrections and alterations—and be aware that an extra round of review could take several months.

When learning how to write a good scientific article, it's best to try it first on your own. However, input and feedback from your co-authors and your supervisor are essential for presenting your work in the best possible form. When you've given the paper a good read-through, be sure to allow the other authors on the paper (including your supervisor, of course) to read it and provide their comments. Once you've made your revisions and have completed a final draft that you're happy with, you're ready to submit your article.

Submitting Your Paper for Publication

The next step in the publication process is to choose the target journal for your submission. When making this decision, keep in mind the type of readers who'll be interested in your work, the journal's impact factor, and the importance of your research results to the field. While everyone hopes to get their research published in a top-tier journal, that might not be the right place for you—or your article. Journal selection is an art in itself and most researchers will aim to have the highest impact possible for their paper. There's nothing wrong with aiming high, but keep in mind that such lofty publications have strict criteria for acceptance.

Another thing you'll need to consider is whether you want to submit your work to an Open Access journal (as opposed to a traditional, subscription-based journal), in which articles are published online and made freely available to everyone at no cost. Similar to subscription-based journals, an open-access article will go through the peer review process. What differs, however, is that once accepted, the article is immediately published online—an added bonus if you want to get your research out in the world more quickly. In most cases, copyright is retained by the author, and there are few restrictions for subsequent distribution or reuse once the article has been published. Some subscription-based journals do publish selected articles as free content, though the journals are not in themselves considered to be open access.

As a PhD candidate looking to grow your career, what does open access offer you? The unrestricted distribution of your published article means more people will read your work. In addition, as a researcher, you can access and create an online repository of the most recent (open access) research in your field.

Whether open access or subscription-based, once you've decided which journal you'd like to submit to (with one or two backup options if your paper is rejected), it's time to move on to the next phase of editing: polishing your prose and scrupulously abiding by the journal's Author Guidelines.

After you've created a version of the text that's been agreed upon by all co-authors, and approved by your supervisor, you're ready for the next stage: the publication process. The art of getting a paper published could take up an entire book on its own; instead of writing a separate tome, we've summarised for you the key steps of the process.

> Take a look at **Chap.** 15 for tips and suggestions on remote collaborations with your co-authors.

As the saying goes, 'you won't get a second chance to make a first impression,' so at this stage of the game, it's important to anticipate what editors—and reviewers—are looking for in a quality manuscript, and to pinpoint anything that might annoy or confuse them about yours.

For a non-biased opinion, ask a fellow PhD student or Post-doc in your department to review your manuscript before sending it out. That individual should be familiar—but not too familiar—with your field. Be sure to return the favour to others so you can learn the fine art of reading manuscripts through the eyes of a reviewer. Journal selection is an art in itself. It's natural to want to publish your research in a high-impact journal. Or perhaps you dream of publishing in the most prestigious journal in your field. However, high-impact journals have stringent criteria for acceptance and the field of potential authors is crowded.

Before submitting your paper to the target journal, it's a good idea to store a copy of your manuscript in a central (online) archive, such as Arxiv. Once you press the 'submit' button, the first step in the publishing process is that your paper lands in the editor's inbox, after which it may be rejected straight away without review. To avoid an immediate rejection, and increase your chances of publication, we suggest choosing a journal where you'll have a good chance of receiving a proper in-depth review (so you can learn from the feedback).

While it may take some time to get the review back (even after a gentle query), it will most likely be a mixed review. Different reviewers will have different perspectives, of course, with each reviewer providing positive and negative comments. Be thankful for the positive comments and expand on the feedback as needed. With any negative feedback, be prepared to make compromises and to find room in your arguments for improvement. If any inconsistencies or errors have been noted, make the necessary corrections to the text, figures, and tables.

Finally, resubmit the revised manuscript.

If your manuscript is ultimately rejected by the editor, you'll feel disappointed, of course, but it's best to move on. Adjust the article as needed to fit the layout and requirements of your second-choice journal and try again. Be aware, however, that—according to some analyses—it can take an average of three months (or more) for the entire publication process, from submission, review, acceptance, and publication.

Closing Ceremony

Congratulations! Your manuscript has been published in the journal of your choice. Wherever you happen to be on the science career ladder, a published article is always cause for celebration. For ideas on how to celebrate your success, see Chap. 17.

Are Traditional Scientific Journals and Publishing Practices on the Road to Extinction?

The digital revolution sweeping the globe for the past 20+ years has had an enormous effect on the way we create and consume information. From the introduction of Amazon's Kindle in 2007 to the advent of smartphones and countless apps for online reading, listening, and sharing of information, many people naturally wonder if printed books and journals will eventually suffer the same fate as the dodo and *Tyrannosaurus rex*. Though to be fair, the demise of the dinosaurs was hastened along by an asteroid strike, and printed journals still fill our libraries and bookshelves—for now. In a rapidly changing world, there are several drawbacks to the current model, however, including the issue of 'publication bias'. This refers to the tendency of journal editors to preferentially publish research with positive or ground-breaking results, which could increase the risk of authors inflating the significance of their findings or overly hyping a new experimental approach. Some fields of science are experimenting with the use of online notebooks to publish their research, rather than go through the more traditional—and cumbersome—process used by peer-reviewed journals, along with their typically long lead times to publication. Proponents argue that it's better to present data (accessible to all) as a type of living document, rather than a printed (or online) journal article that quickly becomes dated or gathers dust on a library shelf. Knowledge and information exchange are accelerating quickly, helped along by online publishing and search algorithms. We're not quite ready to say good-bye to print journals yet, but the long, slow demise has perhaps already begun. Over coffee, Peter and Isabel discuss the pros and cons of shaking up the traditional publication process. As a chemist, Isabel relies on hard data to understand the research in her field and to guide her own experiments. She's worried that ever-evolving online notebooks and data that are published without the benefit of peer review could lead to a 'Wild West' situation, with few rules, and where anything goes. Peter, as an art historian in training, likes the idea of online collaborative publishing, since accepted paradigms can quickly change based on new evidence and discoveries. Whether published online or in print journals, he worries that the lengthy peer review process and gatekeeping by journal editors could mean that a lot of interesting work is being kept from a wider audience.

17

Celebrate Your Success

Success is not the key to happiness. Happiness is the key to success. If you love what you are doing, you will be successful.
—Albert Schweitzer

At long last you have obtained the results you've been working towards for such a long time. Hurrah! You may be so busy (and tired) that you may not even realise that you have indeed achieved an important measure of success. Perhaps it will take a few more months before you can present your results at a conference or submit an article to a scientific journal, but what you ultimately present or submit will be based on the results you've just obtained. Congratulations, you have reached an important milestone, so it's time to celebrate! Too often, success is not celebrated properly, and you simply return your 'nose to the grindstone' without even taking a moment to pat yourself on the back. In this chapter, we make an argument for the importance of celebrating your success, as well as taking the time to thank others for their contribution and support. So, whatever you do, put down your lab notebook, turn off the Bunsen burner, and take a moment to bask in the glory of the rewards of hard work.

The Art of Celebrating Success

Striving for the best requires a great deal of effort, and it's likely you'll encounter many hurdles on your way to the top. Such platitudes hold true for

P. Gosling and B. Noordam, *Mastering Your PhD*, https://doi.org/10.1007/978-3-031-11417-5_17

many areas of life—including scientific research and high-level sports. There is, however, one major difference between athletes and researchers. In the sporting world, athletes seem to know how to celebrate their achievements. No matter what you might know about the sporting world (or whether you care very much about it), you're probably familiar with the way athletes celebrate their triumphs. We have all seen pictures of overflowing champagne bottles or athletes cavorting with glee as they cross the finish line ahead of the pack. Many sports have their own traditions for celebrating a big win.

The world of research seems to lack the tradition of throwing a proper party. We are serious scientists, after all! But sometimes it's necessary to 'let your hair down' and celebrate your achievements.

Why Celebrate Your Success?

Scientific research can be a long and tedious process. It starts with ideas and brainstorming, followed by research protocols and experiments, and ends with a report to the scientific community. But it shouldn't stop there. Here are three reasons why proper celebrations should be an integral part of the research life.

1. **To acknowledge co-workers for their contribution to your success.** We make progress in life, and work, because we stand on the shoulders of others. It goes without saying that we make use of collaborative networks established by others; we use equipment built and designed by others; we analyse data using software written by others; we do research based on concepts sketched by others, and use questionnaires developed or validated by others, and so on. In spite of this, however, it is natural to feel that your work (and your work alone) made your recent progress possible. We all have a tendency to underestimate the contributions of others. By celebrating your success with friends and colleagues, you acknowledge, in an explicit way, their contribution to your success. They deserve the recognition, and by thanking them in a visible way, they will likely be more willing to help you again towards your next milestone.

2. **Because reflection is an important part of the learning process.** You've probably discovered that you can learn valuable lessons about a process by studying what went wrong. But it is equally important to reflect on your successes. Why did it work (against all odds)? Why has nobody else performed these experiments? What triggered the research? What helped you in obtaining the data first, before anyone else? Whose assistance has been critical? Analysing the reasons for your success might help you in the next phase of your research. Luck is for those who know how to find it. And that's something you can learn.

3. **Celebrations create a positive atmosphere.** When you and your co-workers celebrate progress on a regular basis, you will create a winners' mood within the team. In such an atmosphere your team will find more inspiration to tackle the next problem, helping to pave the way to the next milestone.

What Defines Success?

Of course, winning the Nobel Prize is a good reason to throw a little party. But that should not be the standard definition of success. After all, few people actually ever win the Nobel. But during your PhD years there will certainly be a couple of occasions for thanking others for their contribution and support. A very natural moment to celebrate your success is the acceptance of a manuscript by a scientific journal. At some institutes, it is a tradition for the first author to bring cake for the whole team on such occasions. In other research programmes it might take much longer before the publications appear, for instance, because new equipment or methodology has to be developed and tested. In that case you might celebrate when the equipment is ready (don't forget to include the people from the technical workshop) and fully operational.

How Can You Celebrate Your Success?

Celebrating success works the same way as giving someone a thank you gift. It's important that you do it immediately and with the best intentions. Just as you would give someone a nicely wrapped present the day after they pass an exam, you might bring a cake and other celebratory foods to the lab the day after you obtained the key data for your article. Whatever you do, use your imagination and do something fun for yourself and those around you. Everybody enjoys a good party. So take a moment to raise your glass and toast yourself and those around you for a job well done.

Saving an Old Master Painting: The Team's First Joint Article Was Accepted by a Prestigious Journal

Isabel, Yousef, and Peter have each published individual contributions to journal articles in the first years of their project. It was only very recently, however, that they achieved two milestones at once. They wrote their first paper together—with several contributing co-authors and supervising professors—and the article was accepted for publication in a prestigious journal: a significant success that deserves proper celebration.

However, this turned out to be easier said than done. Yousef decided to join his world-renowned professor on his sabbatical and was currently working abroad on a new modelling framework. Peter was staying with his family to support his parents who were faced with some health issues; thus, as a result, only Isabel was on campus. To celebrate their success together nevertheless, they had a small Zoom celebration that included a virtual breakfast. It turned out to be a joyful event in which they learned about each other's breakfast habits, ranging from croissants with a small espresso to herbal tea and fruit salads.

While celebrating, the three discussed how they could additionally boost their success, and the idea of making a short film crossed their minds. Of course, a full movie for regional television would be fantastic, but they soon realised their findings were probably more interesting to a smaller audience. Peter had some experience in video editing and suggested making a 15-minute film to present their inspiring research project. Yousef recorded a three-slide PowerPoint presentation in which he introduced the project. Isabel gave a virtual walking tour of the lab that she recorded on her phone, and Peter summarised the outcome of their project by showing an animation of his modelling of the paint degradation.

They shared their short film with the coordinator of the cross-disciplinary programme for Master's students, who was happy to post it on the university website. Even though the film wasn't aired on television, it received quite a lot of attention: hundreds of views, dozens of likes, and a few enthusiastic comments. Moreover, the short film showing how the three PhD students love what they do motivated a Master's student to join their team for her thesis project. A new team member to support their research project was the biggest gift they could wish for, and an additional bonus to celebrating their success.

18

How to Make the Most of Your Annual Evaluation

If you don't learn from your mistakes, there's no sense in making them.
—Anonymous

One way or another, no matter which programme or department you're in, you will most likely receive an annual evaluation from your supervisor. This chapter will help you prepare for—and survive—that all-important evaluation, as well as give you some advice in getting the most out of what can be a stressful situation.

Very often the yearly evaluation is a requirement orchestrated by administrators high up in the system. Even your supervisor cannot do much about it. You may even believe that it is a waste of time and just another bureaucratic hoop to jump through. But if you approach it in a positive frame of mind, you might be able to see the benefits of this kind of yearly assessment, as it will likely help you come to grips with your progress and performance in the lab.

There is no typical type of evaluation. Although some universities and institutes have standard forms and procedures that must be followed, most likely these forms are rarely used properly, and the official procedures are not followed to the letter. At other institutes, there may be no formal annual evaluation at all. However, every now and then, even in the absence of formal procedures, you should arrange to have a conversation with your supervisor about the long-term perspectives and goals of your PhD programme.

P. Gosling and B. Noordam, *Mastering Your PhD*, https://doi.org/10.1007/978-3-031-11417-5_18

No matter how casual these discussions might appear, they are very important, as the roadmap for your PhD and your performance in the lab will be discussed. In the suggestions given throughout this chapter we have assumed that you will have some sort of a formal discussion with your supervisor on a yearly basis. But most of the suggestions still apply in situations where the discussions are more informal.

Keep in mind that many—if not most—supervisors probably dislike these evaluations just as much as you do. First of all, your supervisor is a scientist, not a trained human resources manager. Most scientists value freedom very highly, which is a perk of academic research, and they may dislike the paperwork associated with universities and government organisations, as well as the type of structure and benchmark-style assessments they believe is typical of the corporate world. Second, as a result of such an evaluation, mutual commitments will be made. For instance, agreements about the type of research you will be doing, and the frequency of your progress meetings, etc. Your supervisor may feel that all these additional commitments will absorb his or her last bit of spare time for doing research, on top of teaching obligations and writing grant proposals. This state of affairs has a way of raising the stakes on both sides of the table, so it may help to keep in mind that the evaluation is not just about you, but will also involve commitments of time and energy from your supervisor. Finally, during the annual evaluation you will very likely receive criticism about your work and your performance in the lab. Most supervisors have not been trained in giving constructive criticism, so you may come away from the experience thinking that the feedback you received was unduly harsh and not constructive at all! Nobody likes to be criticised, so try to keep in mind that your supervisor may be just as uncomfortable as you are. After all, he or she is only human, and your supervisor's opinion is certainly not the last word about who you are as a person or in determining the value of your work and contribution in the lab.

Some supervisors announce the evaluation a few days in advance, as they should do in the spirit of fairness. Others do not. Regardless, you should take the time to prepare yourself for this discussion. If your supervisor does not make a habit of announcing the evaluation, or there is no formal evaluation requirement at your institute, it's a good idea to prepare for an evaluation anyway, so you'll be ready no matter what happens.

First and foremost, if you come prepared to an evaluation session, you'll feel relaxed and confident. Make a list of everything you've accomplished in the past year. Experiments carried out, coursework completed, skills learned, students supervised, classes taught, etc. Then make a list of the areas in which you think you could have done better. This will let your supervisor see that you have spent time thinking about areas in which you can improve. If you don't admit to these things yourself, before your supervisor points them out to you, it will be difficult to avoid the all too human reaction of becoming defensive and inflexible in the face of criticism.

When you sit down with your supervisor, take the lead by presenting an outline of your accomplishments of the past year (make a copy of your list for your supervisor to refer to during the meeting). This way, you start out on a positive note and bring to your supervisor's attention the fact that you have done quite a lot during the year. A list of accomplishments will erase the idea from your supervisor's mind that the project is going nowhere. Second, while discussing last year's progress, be sure to name a number of hurdles that prevented you from making even more progress (e.g., data were not available, equipment broke down, collaborating individuals haven't delivered what they promised, etc.), but be careful not to sound like you're making excuses. It's easy to fall into the trap of blaming outside circumstances and other people's failings for falling short of your goals.

Finally, discussing what should be done in the coming year will establish a road map that will take you closer towards your PhD. Make sure that your plans are not too ambitious and recognise that next year's progress will probably be of the same order of magnitude as the previous year's accomplishments. Making a workable plan for the coming year, and sticking to what has been agreed, will be important for maintaining a good relationship with your supervisor. Not to mention managing his or her expectations by keeping a firm hold on what can realistically be accomplished. Be an active partner in the process, not a passive participant.

In addition to planning, evaluations should be about the things that have gone well—and about things that haven't. Frequently the first part is forgotten, and only negative issues are discussed. So, don't forget to make that list of the most important things you have accomplished. Moreover, thank your supervisor for the things you appreciated about his or her role in your work. Compliments can work wonders (as long as they are sincerely expressed), and by stressing the pleasant qualities and work habits of your supervisor, you will motivate him or her to keep behaving that way. Next on your list should be a few things you are struggling with, some of which your supervisor probably hasn't noticed yet. These can be technical problems, but also social concerns, in fact, anything that stops you from doing your best work. Remember: addressing a problem is halfway towards a solution.

You might believe that the conversation taking place during the evaluation is not in your hands. However, you can control the evaluation in many ways. You don't have to wait until your supervisor comes to you to discuss your long-term plans and last year's progress. Go to your supervisor and tell them you have been thinking about your recent progress, are wondering what to do next, and would like to make an appointment to discuss this. Almost every supervisor will welcome such a pro-active PhD student. During the meeting you will want to make sure that both of you have constructive intentions.

How can you ensure that the evaluation has a long-lasting impact? Once you've gone through what may have been an uncomfortable process, you'll want to make sure that the impact of the evaluation does not fade away the next day. Therefore, you should focus on a few topics and assure that by the end of the meeting you have decided on actionable conclusions to these topics about which you both agree. In the absence of official forms you might formalise these conclusions by sending your supervisor an e-mail in which you list the agreed upon actions. '*We should talk more often*' is not a truly actionable statement. '*Let's sit together every Friday for half an hour after lunch*' is a much better actionable response to your evaluation (and don't forget to make a SMART plan as discussed in Chap. 3).

Altogether, an evaluation is not as bad as it might seem. When properly prepared and carried out, with positive intentions from your side and closed with actionable conclusions, you will get much more out of it than you probably expected at first.

The Surprise Attack: How to Act When You're Caught Off Guard

It might happen at some point that you're taken by surprise. Out of the blue, your supervisor comes to you with a list of things that have gone wrong—or at least that is his or her interpretation. For you, this criticism is all new; you never received even so much as a sign that things might be going wrong. The annual evaluation has suddenly turned into an unpleasant attack on your abilities and performance in the lab. Perhaps you'll react by becoming angry and defensive during the ensuing discussion. Why has my supervisor never told me this before? Why have they failed to highlight the things that have gone well? The way they talk to me makes me feel like a failure.

What can you do about these unfair allegations? Do not expect that the problem between you and your supervisor will be solved during the evaluation itself. It is not very likely that you will convince your supervisor on the spot that he or she is wrong. Neither will the problem go away by just ignoring it. Probably the best thing to do is: (1) try to summarise the criticism, (2) agree to disagree, and (3) ask for a follow-up meeting in a week or so. In the meantime you should prepare for that follow-up discussion. You might even want to talk about the situation with a friend or one of your colleagues. At the follow-up meeting, the following three things should be discussed:

1. **You and your supervisor have a communication problem.** Somehow your supervisor has not been able to communicate your shortcomings during the year and has bottled up and kept to himself all that has gone wrong with your project. It might very well be that your supervisor feels that this yearly evaluation was exactly the right occasion to tell you the truth, with no holds barred. Someone in such an aggressive mood will probably not be willing to listen to counter arguments, so you need to save them for a later discussion. At the follow-up discussion, you need to state the fact that you have a communication problem and discuss ways to solve it. In all likelihood, you've had no proper monthly progress review (see Chap. 6). You might suggest to your supervisor that you start having monthly discussions. Mention that you're willing to prepare the homework for these meetings by filling in the Monthly Progress Monitor. Be pro-active and offer to do the work to take some of the burden off your supervisor, but make it clear that you require more regular discussions in order to improve communication between the two of you.

2. **Establish the things that have gone well.** During the initial discussion, all the attention was focused on things that have gone wrong. Thus, it's important to collect yourself and rebuild a foundation of mutual trust with your supervisor. Only on the basis of this renewed trust, can you hope to successfully proceed with your PhD programme. So, for the follow-up meeting prepare a short list of the most relevant things that have gone well. Try to be honest with yourself when making this list; overstating your good qualities will not help establish new common ground with your supervisor.

3. **Come to the follow-up meeting with a few practical suggestions on the most important (perceived) shortcomings in your work as a PhD student.** No doubt you believe that some of the issues raised by your supervisor are either completely irrelevant or are incorrect. For those issues it should be possible to put a positive spin on things and identify ways in which you have made progress. Remember, building common ground to continue your PhD is just as important as resolving the issues you have. You might even admit that some of the issues raised by your supervisor

have some grounding in reality. If you have a plan about how to work on a specific issue, tell your supervisor. If issues were raised that you have no idea how to improve, be honest with your supervisor about this. He or she has a stake in your success and should be willing to help you on those issues; in particular, since you have demonstrated your commitment by coming up with a solution for other issues. Your supervisor will be happier upon realising that telling you the truth, however painful, has indeed resulted in a positive outcome, and inspired both of you to improve the way you communicate with each other.

Evaluating Your First Year

Congratulations! You've completed your first year as a PhD student. Perhaps the time has flown by—or dragged at times of uncertainty and doubt. Being a PhD student is likely a full-time job for you that occupies most of your waking hours. The first few weeks were likely a whirlwind of activity: meeting your new colleagues, getting to know your supervisor, and adjusting to a new environment. Filled with excitement about your research project at this stage of your scientific career, you've probably enjoyed many aspects of your new position, while others might have turned out to be less inspiring or upbeat than expected.

Though you've got several more years to go as a doctoral candidate, by this point, you've likely settled in to enjoy the ride, even while knowing you'll encounter the inevitable bumps and potholes along the way. In time, you'll get used to the rough spots or minor irritations about the role, as your fascination with—and commitment to—your research project grows. On the other side of the coin, it's possible that, faced with the inevitable doubts and setbacks that are typical of first-year PhD students, you may have had moments when you've considered stopping altogether. Perhaps you've already recovered from a major setback and moved on. But what do you do if, during a dark night of the soul (or several), you find yourself plagued by an uncomfortable and persistent thought: '*Is a PhD really for me?*'.

In Chap. 7, we discuss possible strategies for managing such setbacks and doubts. In this chapter, we suggest using your first-year evaluation as a time for reflection. At this point, you've gotten a handle on what this 'getting a PhD business' actually entails. Perhaps, due to setbacks or disappointments, you've missed out on the adrenaline-boosting thrill of success. If so, the good news is that the excitement of success is yet to come. However, earning your doctorate will require many years of hard work and dedication. So, if you're really certain you want to 'disembark from the PhD train', it's better to do it now. You've already experienced a full year of being a graduate student and working on your PhD topic. Spend some quiet time reflecting on your progress thus far. If you come to realise that moving forward is not the right path for you, it's probably best to stop now and avoid the frustration of several more years of working on a project, and towards a degree, that isn't right for you. It's your life. You're in charge, and there's no downside to acknowledging that the path you're on isn't leading you in the right direction.

As part of their role in training new scientists, universities recognise they have an obligation to provide PhD students with a fair evaluation at the end of their first year. Of course, your institute may not want to let you go after investing in you and your project for a year. But both your supervisor and your university have a responsibility to ensure you're on the right track. In addition to your supervisor, some institutes arrange a small panel to provide input. This panel, after meeting with you and reviewing your progress in the first year, may advise you to either discontinue your project or to continue pursuing your PhD.

You might view such an evaluation as a test of sorts, but we suggest you see it as an opportunity to obtain important feedback from other people who have been exactly where you are now, as well as an opportunity to reflect whether it's in your best interest to invest multiple years to reach your original goal of earning a PhD, or to stop early and avoid years of frustration.

How bad would it be to stop after a year? Does it mean (as you may be thinking during a 'dark night of the soul' moment) that you're a failure? Absolutely not! You've had an extended stay in the halls (and laboratories) of academia, became acquainted with the system, learned about the nature of scientific research, what it means to work with a team, and focused on challenging problems. All of these things are beneficial for your professional career, whether you decide to continue on—or to stop after one year. Although bowing out before you attain your degree will exclude you from an academic career (unless you start over elsewhere), most PhDs end up working outside academia, so it just means that you'll enter the vast universe of non-academic work a little bit earlier.

Going ahead with your training, especially after experiencing a setback or disappointment, is a mark of perseverance. Dogged persistence, however, particularly in the face of diminishing returns, has a negative side to it: continuing on a track that brings you little joy, and one you won't likely finish, won't give you the energy and excitement you need to see your PhD project through to the end. So, should you stop altogether? Take some time to consider your options. We have no doubt the right answer will come to you in a moment of clarity, and the path forward will appear through the mist.

Saving an Old Master Painting: Isabel in the Hot Seat

Wednesday morning, just after teaching his classes, Isabel's supervisor drops by her office. After some chitchat about the lack of knowledge of the first-year students in his classes, he suddenly announces that they should talk about the progress of Isabel's project and what she should do in the coming year. Although Isabel very much wants to have such a discussion, she is absolutely not prepared at this moment, which makes her feel uneasy and caught off guard. She manages to suggest that they have their discussion the next morning rather than right away.

That afternoon Isabel thinks about the aspects of her project that are most important and comes up with three[1] items she wants to discuss: (1) she likes the interdisciplinary teamwork with Yousef and Peter and wants to ensure that she can continue to work in that setting. (Remember that confirmation of the things that are going well is an important aspect of the discussion during an evaluation); (2) she wants to attend the large meeting on art restoration next year; and (3) she is hoping to start her new project soon, as she anticipates long lead times in the preparation of samples.

At the meeting on Thursday, her supervisor asks whether Isabel would like to say something first. Having prepared beforehand, she takes the opportunity to do so. The conversation starts on a good note when Isabel mentions that she is really enjoying the teamwork. Concerning the other two issues she raised, her supervisor is not willing to answer immediately. After some polite insistence from Isabel, her supervisor does set clear goals about what should be done to register for the meeting and what needs to be completed before a new project can be started. When Isabel mentions the long lead time of some parts of the new project, her supervisor is pleased to hear that she is 'staying ahead of the problem' and agrees that she should work on the new project for half a day a week.

Towards the end of the meeting her supervisor asks the usual question: Do you have anything else to say? A few minor points cross Isabel's mind but she decides to drop them in favour of a crucial point: that they have agreed to create a short list of conclusions that came out of the meeting, which Isabel later e-mails to her supervisor for his approval. Isabel now has a written record of their conversation and their mutually agreed upon plan of action.

---

[1] **Three**, for practical reasons, is typically a good upper limit for the things you want to discuss in one meeting. Restricting yourself to the three items most important to you is always a good strategy. Few people can successfully defend more than three arguments, or recollect in detail more than three complex issues at a time.

19

The Final Year: Countdown to Your Thesis Defence

Education is not the filling of a pail, but the lighting of a fire.
—*W. B. Yeats*

You're nearly there. One or more articles are already in print, and another one is in the publishing pipeline. Moreover, you have masses of promising data that still need to be analysed. It's not yet clear, however, how these analyses might be integrated to create a cohesive body of work, so there's still some planning to do. Finally, there is one last project your thesis advisor wants you to work on. Altogether, there are several bits and pieces that fit into your thesis, and other things that don't seem to have a logical place. You're feeling confident that you'll be able to pull everything together within a year's time. But at the back of your mind a lingering question arises: how do I get there? This chapter provides you with a framework that can support you in the crucial last months of your doctoral work.

Make a List of Your Achievements

The first step in planning the route towards your final goal is often neglected as unimportant, but it's essential to get a clear picture of where you are right now. To make explicit what you've achieved so far, you'll need to make a list of two types of research projects.

Start with the projects that you've completed. For example, a part of your research that has already been published as a journal article and needs only

© The Author(s), under exclusive license to Springer Nature Switzerland AG 2022
P. Gosling and B. Noordam, *Mastering Your PhD*,
https://doi.org/10.1007/978-3-031-11417-5_19

a bit of editing to fit in your thesis. Next, list the unfinished projects you're currently working on, and try to identify the steps you'll need to take to finish those projects. These might include tasks such as sorting out the relevant data, writing and running a data analysis programme, summarising the key results from the analyses, or searching the literature (again) to find out if any conclusions that are closely related to your own research, have already been reported. Don't forget to factor in any potential hurdles you'll have to overcome to finish your ongoing projects.

Verify Your Achievements with Your Supervisor

Now that you have a good idea of what you've accomplished thus far, it's time to make an appointment to discuss your insights with your supervisor. She or he might have a slightly different perspective on what you've achieved. For example, your supervisor may feel that not all of your finished research projects are appropriate to include in your thesis, or perhaps they will suggest that you include in your thesis some research that you have done in collaboration with others. Concerning the ongoing projects, your supervisor will probably not have the same detailed insight in their status as you do. He may underestimate what still needs to be done or bring up additional aspects of the projects you haven't thought of. So, ask for a meeting with your supervisor to discuss all the results you have achieved so far. Submit your analysis in the form of a brief written report with your supervisor prior to your meeting. This will hopefully motivate your supervisor to prepare for the meeting ahead of time and to reflect on the status of your research results. The key outcome of such a meeting is to establish a common understanding of what you've achieved for your thesis so far, and what still needs to be done.

What Else Should Be Included in Your Thesis?

Most likely, you and your supervisor will agree (after a productive conversation) about what has been accomplished thus far, and what still needs to be done to complete your ongoing projects. Opinions might diverge, however, when it comes to additional material that should be included in your thesis. Sometimes there will be other stakeholders, in addition to you and your supervisor. For instance, it could be that your direct supervisor is not your thesis advisor, or there may be a thesis committee that takes an active involvement in defining the content of your dissertation. For the remainder of this chapter, we will assume you have both a direct supervisor and a thesis advisor.

Consider the following scenario: you believe that completing your ongoing projects will provide sufficient content for a thesis. Your direct supervisor, however, wants you to include a minor addition, whereas your thesis advisor, who has only had limited involvement so far, wants you to start a new ambitious project. Debating and agreeing on these issues will not easy. But you'll need to make up your mind about what you believe is a proper balance between the quality and quantity of the content, on one hand, and the time it will take to get there on the other. It's best to make this decision before you start discussing these issues with your thesis advisor. Preparation for this discussion is a two-step process: first, describe the final (still-to-be-completed) project that needs to be included in your thesis; and second, make a countdown list (see below).

Describe Your Final Project

To facilitate your discussion with your supervisors on what remains to be completed, your brief description of the (potential) final project should address at least three issues: first, prepare a brief project description, including the means of acquiring the data, your plans on how to analyse the data, and your best guess on the type of answer you might get from your research, including the hypothesis you're testing. Second, you should make an estimate about the types of hurdles and pitfalls you might encounter and how you plan to overcome them. Base your estimates on the experience you gained in your previous projects. Once you know the required actions to finish the project, you'll be able to estimate the time it will take to finish it. Again, do not be too optimistic or unrealistic. It is best to base your estimate on your past performance in similar projects. This exercise will provide you with the missing information you need to complete your countdown list.

The Countdown List

An example of a typical countdown list is shown in the inset and consists of two parts. The first part describes the content of your thesis (your ultimate goal), while the second part describes the planning on how to get there. Writing down most of the chapter titles should be straightforward, once you've had a proper discussion with your supervisor about your achievements and results to date. The chapter on the new project may be an open-ended issue. By including that chapter both in the content list and planning section, it will become clear what the consequences will be if you add this project to your thesis. In the countdown list, time is allocated to each task. In reality, you won't be executing one task at the time, but rather working on a few tasks in parallel. To get an impression of the total time the thesis preparation will take, you can just write down the tasks sequentially. The best way to estimate how long a particular task will take is to rely on your own experience of executing similar tasks. Otherwise, consider consulting your peers.

The Countdown List

Part I—Contents of your thesis

Chapter 1:   Introduction to the field of research.
Chapter 2:   Methodology, research instruments.
Chapter 3:   Research project 1 (already published as a journal article).
Chapter 4:   Research project 2 (manuscript submitted for publication).
Chapter 5:   Research project 3 (data available, no conclusions drawn yet).
Chapter 6:   Research project 4 (the new project?).
Chapter 7:   Conclusions/summary.
                 List of references.

Part II—Planning

(Note: the order of these actions might be different in practice).

Week 1	Transform your journal article into Chap. 3.
Week 2	Transform the submitted manuscript into Chap. 4.
Week 3–4	Write introduction (Chap. 1).
Week 5–6	Write the methods section (Chap. 2).
Week 7–10	Analyse the data for Chap. 5.
Week 11	Conference preparation
Week 12	Conference
Week 13–14	Draw conclusions for Chap. 5.
Week 15–17	Prepare final project (Chap. 6).
Week 18–20	Write Chap. 5.
Week 21–22	Vacation.
Week 23–29	Data acquisition for Chap. 6.
Week 30–33	Analyse the data for Chap. 6.
Week 34–35	Draw conclusions for Chap. 6.
Week 36	Start job search.
Week 37–39	Write Chap. 6.
Week 40	Write summary for Chap. 7.
Week 41–43	Buffer (at least 5–10%).

Week 44	Finalise draft thesis for the thesis committee (*).
Week 45	Approval of thesis (*).
Week 46	Printing of final version of your thesis.
Week 47–48	Continue job search (see Chaps. 22, 23 and 24).
Week 49	Prepare thesis defence.
Week 50	Thesis defence (*).

(*) Note that the actual procedure for the final approval of your dissertation and thesis defence differs between universities. Check with your thesis advisor about how much time it typically takes at your institute.

Typical Things You Might Learn from Making the Countdown List

Once you've made such a list, you should take a moment to reflect on its consequences. The first thing you'll discover is that time flies. You may have already made a mental estimate of how long it will take you to finish your thesis, but now that you've made your list explicit by writing it down, you may notice it will likely take longer than you thought. In addition, tasks you may have ignored initially, but are now clearly important, will also eat into your time frame. But don't panic. This is the moment to hone your time management skills, and your ability to focus on what is important. Your schedule is tight, so good planning is critical. While you can control your own time management, be sure to take the necessary measures to ensure that any delays, caused by delayed input from others, will be minimal. If you're working in a team and need crucial data from the others, now is the time to plan a meeting to discuss exactly what needs to be accomplished in the (limited) time that is left. Finally, towards the end of this period, you will have to deal with the often-complex procedures that come with submitting your thesis for approval. Make sure you know well in advance exactly what needs to be done and when.

Handling Uncertainties

Probably the biggest uncertainty in your planning scheme will be the time you need to allocate to execute your new project (7 weeks, in our example). Note, however, that if you exceed the allocated time by 4 weeks, the total planning (of 50 weeks) changes by less than 10%. Your total margin of error could indeed be larger. However, you might be in the unfortunate situation that the time it takes to execute one of your tasks is very uncertain. For instance, in the new project you'll have to rely on a sample that needs to be prepared by other people. Without that sample the project will be impossible to complete. If you foresee such problems and other potential bottlenecks, it may be useful to make two countdown lists. One, assuming that the sample will be delivered to you on time, and another one in which that particular project is impossible to complete (see also inset). Estimating these risks upfront will enable you to have a valuable discussion with your supervisor about these potential scenarios. Most importantly, this type of forward planning will minimise the uncomfortable feeling that you have a potentially critical problem ahead of you that is completely unanticipated and out of your control.

Discuss Your Planning with Your Supervisor(s)

Now comes the hard part. You'll need to discuss your planning with your supervisor and thesis advisor, although you might feel that this is more a negotiation than a discussion. Seated at one end of the table is you, the hard-working PhD student, who wants to wrap up the work in a reasonable amount of time (you don't want to be in graduate school forever!). At the other end of the table is your thesis advisor, hungry for more research results that can be used in a future presentation or publication. Your daily supervisor is placed somewhere in between. However, you'll be well prepared to back up your arguments by having made the countdown list. Perhaps your supervisors are not skilled communicators—they have been trained as scientists, after all. But your carefully structured plan will likely appeal to them. Give them your plan well before the meeting so they have time to digest it. In the event that you have a major disagreement during the meeting, keep your cool and avoid becoming irritated. Summarise the issues you don't agree with and ask for some time to reflect on their point of view. Sometimes these planning discussions can't be finished in a single meeting, but they're worth doing properly. You'll save yourself quite a bit of thesis preparation stress if you manage to structure a countdown plan that all parties can agree on.

Your Countdown Planning is not Carved in Stone

Once you've created a countdown list that your supervisor and thesis advisor have agreed upon, you should try to stick to it. However, while working on the tasks in the countdown list you may notice that you're deviating from the planning. Even worse, gaining back lost time is not a likely prospect. If you find that you're really getting off track or that things are taking much longer than anticipated, your first reaction may be to work even harder and longer hours, or to go into denial about how much time you really need to finish. Rather than sticking your head into the sand, face your situation, make a new plan and discuss it again with your supervisor.

Saving an Old Master Painting: Yousef Wraps up His Final Year

Yousef is in the last year of his PhD programme. From the beginning, he has explored a new line of research in a small, start-up group. While attending a prestigious international conference, his supervisor reported on the lab's plans for a new ambitious project. Although some results have been obtained, a major breakthrough has yet to occur. In particular, they've had difficulty preparing the high-quality samples required for this type of research. Yousef feels uncomfortable about the situation, and at this late stage, finishing in time seems impossible. Even if he gets the six-month extension he's applied for, he has just over a year to complete his thesis research, not to mention writing and submitting his thesis for approval. After taking these factors into account, Yousef realises he only has a few months left to do the actual research. Some of his peers suggest making a countdown list. But even though Yousef, an ENTJ (see Chap. 8), is a strong believer in lists and structures, he's afraid that a list will be of little help. The uncertainty of finishing the new project, in particular obtaining the high-quality samples he needs, goes far beyond the planning of tasks on a weekly basis. Yousef ends up spending more time thinking about what to do than he does actually working on the project.

His supervisor, Paul, notices the slow-down in progress and spots Yousef hanging around in the hallway chatting with colleagues, rather than working in the lab. On several occasions, his supervisor passes by and urges Yousef to speed up the work, reminding him of the deadlines that are looming. This passing communication between them is of little help, however. Yousef decides to talk to his mentor in the department. The mentor suggests that Yousef talk to his supervisor about establishing a Plan B, one that involves some simple experiments, just in case the more ambitious project fails. Yousef and his mentor realise that the supervisor is probably not yet willing to give up on the project, because if it's successful, it will bring so much glory to the newly started group—and to his supervisor's reputation in the field. Therefore, the mentor suggests that Yousef do two things. First, convince his supervisor that a 'Plan B' is the only way to rescue his thesis. Second, a moment should be defined when it's time to switch to Plan B. To structure the discussion, Yousef created two scenarios, as described in the countdown list. In the first plan, he assumes that the much-desired breakthrough occurs in a reasonable amount of time (e.g., 3 months). The second plan describes how to move forward if, despite his efforts, the sample preparations don't work out. When Yousef mentions the two options for moving forward, his ambitious supervisor is upset that one of the options would mean abandoning the project. Wisely, Yousef decides not to reveal the countdown plans just yet. A few days later, Yousef and his supervisor meet at the coffee machine. His supervisor admits the project may be a little too ambitious for Yousef to successfully complete. Yousef finds it difficult to accept this subtle attack on his skills, when his supervisor's confidence about the feasibility of the project's success has also played a role. Yousef suggests that they talk again. At that second meeting he brings the countdown plans and shows them to his supervisor. After some discussion, they agree to do the utmost in the coming ten weeks to make the project a success. If, by that time, the sample preparation still hasn't worked they will switch to Plan B. Yousef feels relieved. Now he can start concentrating on his work. Much to Paul's surprise, he rarely spots Yousef hanging around in the hall anymore. Yousef is much too busy in the lab working towards the completion of this thesis research.

20

Writing Your Doctoral Thesis with Style

If I have seen farther than other men, it is because I stood on the shoulders of giants.
—Isaac Newton

All the raw material for your thesis is ready—at least it should be if you've successfully worked your way through the countdown plan described in Chap. 19. Now it's time to wrap things up and write your doctoral dissertation. This is easier said than done, of course, and you cannot hide from the fact that you still have a lot of work ahead of you. But don't let the thought of writing your thesis paralyse you into a state of inertia. Like any big project, writing your thesis is easily doable if it's broken down into smaller steps. In fact, you have already done this by having written at least a couple of articles that are ready to be transformed into chapters. Keep in mind, however, that a research article written for a peer-reviewed journal is not the same thing as a chapter in your thesis. Even if you've published several articles, you can't just staple them together and—voila!—create an instant doctoral thesis. There are some fundamental differences in these two types of writing which we will discuss in detail in this chapter.

Although you may be feeling a bit stressed (or even a lot stressed) that the deadline for completing your thesis is approaching, writing your dissertation will be a new (and hopefully exciting) phase of graduate school. See it as a challenge, and whatever you do, don't stop now. Too many graduate students leave their programme without having written their thesis. One way or the other, your degree will be of help in your further career. It would be a shame to have done so much work, only to exit your doctoral training without that

© The Author(s), under exclusive license to Springer Nature Switzerland AG 2022
P. Gosling and B. Noordam, *Mastering Your PhD*,
https://doi.org/10.1007/978-3-031-11417-5_20

coveted degree in hand. Likewise, you may be the kind of person who loves working in the lab but hates sitting down to write. In this chapter we present a few suggestions for making the process as straightforward and painless as possible.

First Things First: Decide on the Table of Contents

If you didn't make a count-down plan as described in the previous chapter, there is at least one aspect of it that you need to address before you start writing: make sure you and your supervisor agree on the table of contents of your thesis. This might seem obvious, but we have seen too many students start working on chapters without discussing it with their supervisor, only to find that those chapters had to be tossed out. As soon as you have agreed upon the table of contents, you should discuss with your supervisor as soon as possible, and in more detail, about what you are going to put in those chapters. We've repeated here the schematic Table of Contents that was presented in Chap. 19.

Table of Contents

Chapter 1: Introduction to the field of research.
Chapter 2: Methodology, research instruments.
Chapter 3: Research project 1 (already published as a journal article).
Chapter 4: Research project 2 (manuscript submitted for publication).
Chapter 5: Research project 3 (data available, no conclusions yet).
Chapter 6: Research project 4 (data available, not analysed).
Chapter 7: Conclusions/summary.
            List of references.

Cut the Problem Down to Size: Write an Outline

When you're satisfied with your table of contents, it's time for the next step: writing an outline. For a written document as complex as a doctoral thesis, it is essential to work from a solid outline to keep you on track and provide you with a framework for the various sections of text. Writing an outline will also encourage you to break up the writing process into manageable pieces. Your outline should consist of several pages that contain chapter headings, subheadings, figure and table titles, as well as a few keywords and essential comments. Once you've created a comprehensive outline, you will have a framework or scaffold from which to work. In addition, an outline is a great tool for preventing writer's block, as you only need to fill in one section of your outline at a time, rather than face the enormous task of writing a document that will be in excess of a hundred pages. With an outline in hand, each time your hands are poised above the keyboard, your aim is not to write an entire thesis—a daunting prospect, no doubt—but something much simpler. Your new aim is to write a paragraph or section under one of your subheadings. It helps to start with an easy section: this will get you into the habit of writing and give you the self-confidence to write your thesis all the way through to the end.

Getting Down to Fundamentals: What's a PhD Thesis Anyway?

Depending on whom you ask, you'll probably get a different answer to this question. But before you get heavily involved in the writing phase, it may help to get a firm grasp of what it is you're attempting to accomplish. Most people tend to agree on the following definition of what a PhD thesis is (and is not):

A PhD Thesis is

- A formal document, quite substantial in length, that presents original data in support of a particular thesis or supposition.
- A comprehensive body of data that supports a particular hypothesis and is backed up with appropriate evidence. The scientific method requires you to state a hypothesis and then gather data to support or negate your hypothesis. Before you can write a thesis that defends a particular hypothesis or hypotheses, you must gather *sufficient evidence* to support it.
- A thorough analysis and interpretation of the data you have collected. This analysis forms the heart of your thesis.
- A document in which *every statement* is supported by citing either the scientific literature or your own (original) work.
- A document in which every statement must be correct and defensible in a logical and scientific sense. There is no room in your thesis for suppositions and conclusions that you've pulled out of thin air.

A PhD Thesis is not

- *A diary of your days in the lab.* You must be able to present your work in a way that demonstrates your mastery of a given topic. You will not be awarded a PhD just for writing down everything you did in the lab over the course of several years.
- *A collection of published articles.* A PhD thesis is similar to writing a book. While you can take your published papers and turn them into the core of your thesis, the thesis as a whole should be able to stand alone and is coherent in presentation and scope.
- *Written in solitude.* It is important to have other people involved in the thesis-writing process, if for nothing else than for checking your first drafts and proof-reading your final ones. You also need to have a supervisor who will tell you when to stop writing. As the person working on the PhD, you're too closely involved in the process. Therefore, you must ask for expert and third-party advice. Remember that a good thesis, just like any text, is designed for the benefit of the reader. So, try to get several people to read your thesis and listen to their suggestions for improvements.

As you work, be sure to keep the above points in mind. It may also help to read several dissertations from former PhD students in your group or department in order to get a feel for the style and tone.

Choose a Straightforward Format and Layout

You can spend an endless amount of time designing a special format and layout for your thesis. If you're not an expert in desktop publishing, however, we suggest you save yourself a great deal of frustration and time by copying the format of another student whose thesis layout appeals to you. Make sure the format is easy to use, however, as you don't want to find yourself spending many days on learning a new and fancy software programme—at least not at this stage. Once your thesis is sent to the panel or committee for review (a process that usually takes several weeks or months), you might want to consider upgrading your layout. For the time being this should not be your major concern.

Transform (Published) Articles into Thesis Chapters

You most likely have a few articles already in print in peer-reviewed journals or at least submitted for publication. So it may seem like a straightforward matter to transform those articles into thesis chapters. But transforming articles into thesis chapters isn't just a question of copying and pasting the appropriate text. The following are some suggestions for creating cohesive thesis chapters from your published articles or submitted manuscripts.

- First, you will have to rewrite the introduction section of each article to put the relevant chapter into perspective with all the others. There is no reason to repeat in your chapter introduction what you have already explained in your general introduction and literature survey in Chap. 1 of your thesis.

- In addition, the methodology section can be shortened since you have already presented most of that information in the chapter on Materials and Methods. Don't make this section too brief, however, since the reader must be able to read each chapter independently without having to flip back and forth to other chapters for important information.

- Update your references. If your published article is somewhat out of date, you should include the latest literature in your list of references. Moreover, refer to the other chapters of your thesis, where applicable, rather than just referring to your published journal articles. The thesis must be a cohesive piece of work in its own right that can be read and understood without having to refer to additional literature.

- Avoid repeating figures used in preceding chapters. No matter how useful it was to show in each single article a (slightly modified) version of your experimental setup, for example, in a thesis such a series of illustrations is often unnecessary and redundant.

- Adapt the format of your article to that of your thesis. For instance, if you transform a short article or letter into a chapter, insert the headings (Introduction, Results etc.) at the appropriate positions.

- Include sections of text that did not make it into the final version of the article. Space restrictions for your article were determined by the journal editors, so you may have had to sacrifice a couple of interesting paragraphs to meet their requirements. Now's your chance to use these additional paragraphs (and figures, tables, etc.), since they will likely provide a valuable addition to your thesis.

Chapter Two: The First Piece of New Text

Now that you have transformed your published articles into chapters, your thesis is starting to get some heft to it. Although you probably realise that the tough part has yet to be done, take a moment to enjoy your success so far. From now on, you will have to write new text for the remaining chapters and that will slow down your progress quite a bit. Since writing the methodology chapter is relatively straightforward, we suggest you start with that one. You have already written several methods sections for your articles, so you probably won't need much help in making a first draft. Since a thesis has fewer space restrictions, you should take the opportunity to describe some of the details of your work that did not make it in the articles. It is better to err on the side of being too detailed than too little. Be generous to the next generation of researchers. A detailed description of your progress—and failures—in terms of your materials and methods, will save them a lot of time.

The Last Set of Data: Chapter or Article First?

Having worked your way through the initial chapters and written most of your thesis, it's time to tackle your final project. In this particular case, you probably haven't written an article yet and will need to decide whether to write the article first and transform it into a chapter or the other way around. If there is stiff competition in your field to get results published as quickly as possible, your supervisor will probably insist that you write the article first. If this isn't the case, we suggest that you write the chapter first, as this approach will allow you to describe your work, including all the details, from which you can select the appropriate parts for an article later on. While the thesis is out for review with your dissertation committee, you can transform the chapter into an article and submit it to a journal.

The Introduction: The Final Hurdle

A good introduction to your thesis is crucial for putting your work into context, and it is probably the most difficult section of your thesis to write. This is your opportunity to describe your work in a broader perspective, including an explanation of why the research was relevant (i.e., to the scientific community and society in general) in the first place. Although you will probably write this chapter towards the end, you should start thinking about it long before then. During your last year as a PhD student, it's a good idea to create a folder on your laptop in which you collect ideas and article clippings that might be suitable for the introduction. Once you start writing the chapter, you'll have a ready source of ideas, some of which might fit in well, while others may be harder to incorporate. Collecting ideas for your introduction ahead of time will require some discipline, but it will save you from writer's block when faced with writing the text. It can be highly stressful if you have no clue what to write, especially with a deadline hanging over your head like the sword of Damocles. Keeping a folder of ideas will be of help while writing a comprehensive and elegant introduction when the pressure is on.

The Summary

You may be required to write a summary for your thesis, but even if you're not, a good summary is essential, so take the opportunity to write a high-quality one, as this is the one section of your thesis that is sure to be widely read. In just a few pages you'll have to describe the main findings of your thesis research, so it's best to write this part after you have finished all the other chapters. Do not try to describe all your results in the summary, however. If the density of information is too high, it may discourage people from reading further. The last thing you want is for a reader to get over-whelmed by too much detail and put your thesis aside altogether. In addition, be sure to clearly designate the chapter numbers in which specific results are described in detail.

Going for Gold: Aiming for an Error-Free Thesis

Since a doctoral thesis is usually written under serious time constraints, it is difficult, and probably not realistic, to write a thesis that is completely free of typos and other minor errors. Spell-checkers do help, but they have limited use for a document such as a thesis that by definition will contain scientific and technical terms that will not be recognised by the spell-checking software (you can build these into a glossary on your computer, of course, but this takes time). In addition, errors of grammar and syntax are not always highlighted, and minor scientific errors can be easily overlooked. Your goal, of course, is to have the minimum number of errors in your thesis as possible. We suggest you do two things to make this a reality. First, put the manuscript aside for a short while after you've written the first draft. Once you're feeling refreshed and have gained some distance from the material, read it over again with a sharp eye, not for content, but in the guise of a proof-reader who is just looking for errors. Second, you should give a copy of your thesis to one or two trusted peers to read. Find a way to reward them for every error they find as an incentive to go through your thesis with a fine-toothed comb.

Be Generous with Acknowledgements

Some universities allow you to thank and acknowledge co-workers at the end of your thesis. Take that opportunity to whole heartedly thank all those people (don't forget family and friends), including other students, Post-docs, your supervisor, and lab technicians, who have made your work possible.

Ten Tips for Writing a Stress-Free Dissertation

1. Don't save data analysis until the very last minute. Plan ahead.
2. Confirm your table of contents with your supervisor.
3. Create an outline (and stick to it as you write).
4. Don't reinvent the wheel: transform your published articles into thesis chapters.
5. Create a time frame (and deadline) for yourself and stick to it.
6. Find a quiet place to write where you will be free from distractions. The lab is usually not a good place to write a thesis. Work from home or in a quiet place like the library.
7. Assign yourself a number of pages to write each day and stop when you're done. This will prevent you from spending 24 hours a day at the computer, agonising over your progress. When you've written your assigned 4–5 pages, then you're finished for the day. Step away from the computer and do something else.
8. Take plenty of breaks and be sure to spend time with friends and family. Just don't bore them, however, by constantly talking about your thesis and complaining about how hard it is to write.
9. Get some exercise, eat well, and take care of your health.
10. Don't work in utter solitude. This is not the time to turn into a hermit. If other PhD students in your lab or department are writing their theses at the same time, consider creating an informal support group where you can share the stresses of writing a thesis and have people at hand who are willing to read or proof-read certain sections or even the entire manuscript.

21

The Final Act: Defending Your Thesis with Panache

The aim of science is not to open the door to infinite wisdom, but to set a limit to infinite error.
—Bertolt Brecht

At last, the day has arrived, and you are almost ready for the final act. You have completed your magnum opus, submitted it to your thesis committee, and received permission to move onto the next and final step: defending your doctoral thesis. Whether you like it or not, it's 'show time'. You may have been working towards this day for several years, but be aware that defending your thesis will require a different set of skills than you're used to using for your regular research and writing activities. The prospect of this 'oral exam' may seem extremely daunting (not to mention frightening) to you at this time, and you may be wishing you had already passed through the whole thing and were holding your coveted degree in your hands. However, if you think of your thesis defence as a rite of passage, a necessary test of knowledge and competence, and the final challenge you must undergo before you reach your ultimate goal, it will not seem all that insurmountable. If you take the time to prepare for your thesis defence, you will feel strong and confident going into it. In this chapter we offer a few suggestions to help you defend your thesis with confidence and panache.

Depending on where you carried out your PhD research, there will be quite a range of formal procedures and regulations for your actual thesis defence. In some countries, and at some universities, the defence is almost a formality, with no tough questions fired in your direction, and no prospect

© The Author(s), under exclusive license to Springer Nature Switzerland AG 2022
P. Gosling and B. Noordam, *Mastering Your PhD*,
https://doi.org/10.1007/978-3-031-11417-5_21

of failing. If this is the case at your institute, your thesis defence will consist of an hour or two of non-aggressive questioning in front of your friends and relatives. At other places, the candidate is endlessly interrogated behind closed doors by an international committee, and there is a small, but finite, chance that the candidate will not pass. For the remainder of this chapter we assume that you will have to deal with a situation that lies somewhere in between these two scenarios: an oral defence that is open to the public in which serious questions about your thesis will be asked, but the chances you will fail are minimal. Although our advice assumes that you will be participating in a mild form of thesis defence such as this, our suggestions should also be of help to those who must undergo a more rigorous scenario.

In our opinion, there are only three things you need to do to ensure that your thesis defence is successful: prepare, prepare, and…prepare.

Familiarise Yourself with the Formalities

A thesis defence has the characteristics of both an examination and a ceremony. All ceremonies, from PhD defences to weddings, tend to have a set of formal rules that must be followed during the ceremony itself (e.g., standing when the committee enters the room, etc.) and things that must not be done (e.g., address the examiners by their first name, etc.). Since you're probably not familiar with these rules, you will have to pay extra attention to your behaviour, all the while having to answer difficult questions and keep your composure.

Combining these two tasks is not easy and may even require a little 'sleight of hand' to pull it off. One thing that can really help put your mind at ease, in terms of the formalities, is knowing the formal procedure by heart beforehand, so you can focus all your concentration on answering the questions posed to you. To that end, you might want to go to a few thesis defences of your peers prior to your own defence. You will get a feel for how the rules work in real life, and there won't be any surprises during you own thesis defence (at least in terms of rules and procedures). In short: don't go into your thesis defence unprepared. Familiarise yourself beforehand with the rules and regulations, such as how to address the examiners, when to stand and sit down, what the dress code is, and anything else that will be expected of you during the ceremony.

Prepare Yourself Scientifically

There is no doubt that you are the expert on the science you'll be discussing and defending during your thesis defence. After all, you'll be talking about work that has been the focus of your time and attention over the past several years. Do not, however, underestimate the committee's knowledge of your subject. Moreover, in the formal setting of a thesis defence, you have one truly big disadvantage: while your examiners have been able to prepare their questions beforehand, you have to reply to them on the spot. Some of your examiners will be very good at finding a few delicate or controversial issues in your work, and they will certainly question you about them. Remember, it is much easier to ask a difficult and probing question than to answer it on the spot, with hardly a moment to collect your thoughts.

While standing in the spotlight, you may even realise (*quelle horreur!*) that it has been quite some time since you even thought about some of the issues now being pointedly addressed. So, we advise you to read your thesis again, this time with a critical eye and perhaps with a highlighter in hand, in the week or two before your thesis defence in order to refresh your memory about the experimental details, the experimental setups, and the results and conclusions that are described in your thesis. As you read, put yourself in the role of the examiner. What would you ask the writer of this thesis? Where are the trouble spots, the unresolved issues, the shaky conclusions? If you can guess some of the questions you'll be asked beforehand (and prepare the answers), you will be much better prepared for the defence itself.

No matter how well you know your own research, and how well you've prepared beforehand, it is not always easy to phrase the answers properly in public. To improve your skills in responding to all kinds of incisive or round-about questions, we suggest you take part in a fun exercise. Invite a couple of peers from your institute to have dinner at your place. Make sure that you invite both experts in your field (e.g., a Post-doc you worked closely with) and those who are less familiar with your work (the PhD student working in another group down the hall). While you serve and eat dinner (the multi-tasking aspect of the exercise) your guests will ask you questions about your thesis. Some of these questions may trigger you to read a particular part of your thesis again, while other questions will train you to bring your work into the context of an outsider's perspective. No matter what kinds of questions you're asked by your dinner guests, however, you are training yourself to respond right away—and with poise—all the while staying cool and collected as you serve and eat a meal.

Prepare Your Act

Since a thesis defence is a formal ceremony as well as an examination, you will have to act accordingly and play the highly scripted part that is expected of you. It will not be sufficient to give a brief reply to a question while staring at the floor. From experience we can assure you that the examiners will not be pleased if you keep replying to their questions with, 'yes, no, no idea,' and so on. Answering a question properly is a three-step process:

1. First, you'll need to listen to the question carefully. Too often PhD candidates stop listening halfway through a question because they believe they know what the question is all about, or they are so nervous they start preparing the answer while the question is being asked. Sometimes, the real question only comes at the very end of a long exposé (in which the examiner may be trying to show off a bit), so you have to listen carefully the entire time the examiner is speaking. To ensure you maintain your concentration throughout a long monologue, you might want to take notes, or jot down key words as they are spoken by the examiner.

2. In the next step, you should begin your answer by rephrasing the question briefly and politely (remember that it's also a ceremony), such as 'esteemed professor your question on the research described in Chap. 4 addresses the issue of the ageing of paint pigments from an interesting perspective. If I understand your query correctly, you're wondering why…' This rephrasing has a twofold purpose, first to establish whether or not you have understood the question properly. Second, it will give you a moment to collect your thoughts and prepare the best possible answer.

3. In the final step, you should answer the question. This might seem obvious, but too often the candidate makes no serious attempt to answer the question posed and starts in on some related or unrelated tangent or explication to make it appear as if the question were being answered. Some questions may be just too difficult to answer right away, or you may be caught off guard and have no idea how to answer the question that has been posed to you. In this case you have two options. First, you could start talking about the research in the chapter while not giving an answer at all and try to bluff your way through it. A better solution is to admit to the examiners that you probably will not be able to provide the full answer to the question raised, but that you will discuss a few issues that can contribute to finding a proper answer. While the public will not notice

the difference, the experts (and most of the examiners are experts) will understand the distinction between a candidate who is prevaricating and sidestepping the question, and one who makes a real attempt to address the question, as thorny and complex as it may be. The latter behaviour is what the examiners will expect from someone soon to be awarded a doctorate.

Your Physical Condition at the Actual Defence

No matter how well prepared you may be, there is a fair chance that you will be a little nervous—or a lot nervous—depending on how you operate under stress. After all, you have been working towards this point for years, and a great deal is at stake. Each individual reacts differently to upcoming stressful situations. You may or may not have already discovered which strategies work best to help you perform well under nerve-wracking circumstances. You may have had some experience with this before (such as in sitting examinations), but the scale of defending a PhD thesis puts the circumstances of a thesis defence into a different class altogether. So, in the days before your thesis defence, try to find a proper balance between (1) focusing on your research by reading your thesis (again and again) or going for a long and relaxing walk; (2) drinking lots of coffee to activate your brain or imbibing in a cup of herbal tea to relax; and (3) preparing your thesis defence locked up in a room alone (this may work for introvert candidates) or sitting in a café with a couple of friends. Whatever you do in the run-up to your defence that will help you feel more poised and relaxed, try not to develop a completely new strategy at this stage for managing stress. It may be best at this point to do what has worked for you in the past. Get some sleep, go for a walk, eat regular meals, talk to friends. Breathe.

It's possible, even likely, that your defence will have a hybrid format, with some of the audience and some of the examiners in the room, while others are connected via remote channels such as Zoom and Teams. This adds a substantial complication to your act as you have to 'perform on two stages', the physical setting in the room and the virtual performance for those remotely connected. Focus on the live audience first, as this will come most naturally to you and those in the room. Those who are remote will understand they are watching rather than participating in a live event; however, do respond to the examiners on Zoom in an inclusive manner when you address them or refer to their previous questions ('as posed earlier by Professor Smith, the results from our study indicate that…').

Take a look at **Chap.** 15 for tips and suggestions on remote collaborations.

Our advice for getting through your thesis defence with a minimum of discomfort and the best chance for success? Prepare, prepare, prepare, and then just let it go and do your best. We hope you sail through your defence with flying colours.

22

Putting It All Together: A PhD...What's Next?

Science is a wonderful thing if one does not have to earn one's living at it.
—Albert Einstein

You wake up one morning and realise, with a twinge of nervous excitement, that you're six months away from your thesis defence. After spending so many weeks and months at the computer writing your dissertation, it is probably hard to imagine that there is life after grad school. Many graduate students are so intent on getting through their thesis defence, they tend to ignore that after the ceremony, a true milestone in your life, they will have to find a job. Deep down, however, you are aware that you must make some important decisions.

For example, do you want to stay in academia? Or would you rather pursue a career in industry? Should you stay in your own country or do you want to explore new opportunities abroad? These questions are not easy ones, and you will probably not be able to answer any of them overnight. This chapter aims to get you started in the decision-making process and assist you in guiding your thoughts. Two issues will be addressed: Which kind of job suits you best, and how to go about getting your dream job.

Opportunities for a Newly Minted PhD

Once you start thinking about the type of job you'd like to have after your PhD, you will soon discover that the possibilities are overwhelming.

© The Author(s), under exclusive license to Springer Nature
Switzerland AG 2022
P. Gosling and B. Noordam, *Mastering Your PhD*,
https://doi.org/10.1007/978-3-031-11417-5_22

A PhD! - What next?

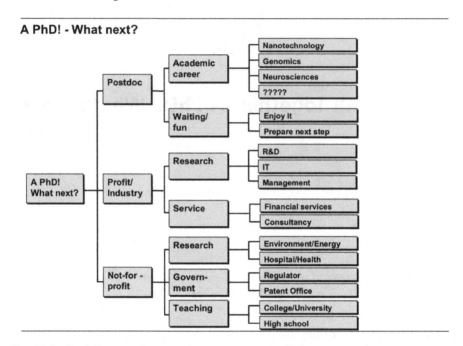

Fig. 22.1 Decision tree for mapping out your post-PhD career options

Somehow, you will have to find a way of making sense of all these potential opportunities—and which will be the right path for you to take. A decision tree, like the one shown in Fig. 22.1, can be of help in narrowing down the choices. This tree assumes you have a PhD in the natural sciences, but keep in mind that it is just an example and, depending on your discipline and interests, may look quite different. This model tree should inspire you to make your own personal decision tree. For many of you the first question you might try to answer is: do I want to do a Post-doc or not? The no-Post-doc category can be further broken down by splitting it into two categories: positions in for-profit organisations and positions working in government and other not-for-profit organisations. In the following sections we discuss these three options in some detail.

The Next Logical Step: A Post-Doctoral Fellowship

You've been training for years as a lab scientist, your supervisor is a scientist, and perhaps a renowned professor you met at a conference has offered you a position in her lab as a scientist. So what's stopping you from taking the natural next step: accepting the offer to work as a Post-doctoral fellow? Post-doc positions are relatively easy to come by and accepting such a position saves you the hassle of going through several job interviews. But there are at least two better reasons than this to continue your career by taking on a Post-doc position: (1) being a Post-doc is the logical and expected next step on the academic ladder; and (2) as it is less stressful than getting a PhD (you've got no impending thesis hanging over your head), it is a good way to really enjoy doing science.

An ambitious PhD with his or her eye on doing a Post-doc, will often wonder which university or research institute is best for them and how to choose a research area that will optimise their chances of working their way to the top of the academic ladder. No doubt it helps to go to a top institute and work in a field that is currently in fashion (e.g., nanotechnology, genomics, neurosciences, quantum computers, artificial intelligence). But there are many other ways to become successful in the long term. Whatever you decide, make sure you do what is right for you, and that you are making decisions that are realistic and will help you reach your goals.

Looking further down the road, when you apply for an assistant professorship (after being a Post-doc for a few years), the search committee will want to see what you have achieved during your post-doctoral period. If the science you have done is impressive, the rest is less relevant. Ask the senior scientists around you what choices they made to get them where they are now. As you will probably discover, they have all done good research, some at top schools, on trendy projects. If you decide to do a post-doctoral fellowship, while knowing that you do not intend to pursue an academic career, make sure you choose a position, location, university, and/or lab that you will enjoy, and start thinking about what you will do afterward, such as work in industry or for a not-for-profit organisation. Also, consider getting involved in some extra-curricular activities that might be relevant to your new field.

For-Profit Organisations: Where Business and Science Meet

We'll start with the most important message first: if you think you want to work in industry, do not wait. Get started on your search right away. Jobs that most PhDs have in for-profit organisations can be classified into two types of industries: (1) research-related companies that hire skilled experts, and (2) service companies that need analytically skilled staff.

Research-related industries either make high-tech products (e.g., Microsoft, Apple, Philips, Siemens, IBM) or need technically trained people to make their products (e.g., Shell, Unilever, Novartis, Roche, Springer-Verlag). These industries tend to be large and multi-facetted. Jobs in small and medium-sized enterprises (SMEs) are scarce these days, but not impossible to find if you have the right connections. Often, scientists start in the IT or R&D branch of the company, and then move on in the course of 5–10 years to management positions. In some cases, you might be able to start immediately in a management-type role. Many newly minted PhDs are put off by the idea that they will not be able to do research forever within a company—after all, isn't bench science what you've been trained to do? However, very often you will find that your interests change over time. For the few real scientific diehards, and those who can't imagine working outside the lab, almost all companies have some senior research positions.

Service companies tend to be less interested in the actual content of your PhD research, but value your analytical and problem-solving skills. Quite a few PhDs find jobs as financial analysts, risk assessment analysts, or management consultants. All major financial institutions and global operating consultancy firms hire talented people with a PhD in the natural sciences, as well as in other disciplines.

Not-for-Profit Organisations—The Best of Both Worlds?

Several years in the lab have taught you one thing: you like doing science and using your analytical skills, but you want to avoid both the insecure academic track and the acutely business-orientated approach of for-profit companies. If this sounds like you, a not-for-profit organisation may offer an appealing career alternative.

Actually, there is a wide variety of jobs that welcome PhDs in the natural sciences, with a broad range of emphasis on doing research. We discuss three of them in decreasing order of their research component.

- **Government labs.** Rather than concentrating on the fundamental aspects of research, government-sponsored laboratories focus on areas of research that have a broad impact on society, such as future energy sources and technologies, public health, and the environment. Their research goals are often long term, and relevant to society's needs. By focusing on longer term horizons, the pace in these laboratories is often less hectic than that in industrial laboratories. The nature of the topics, however, and their relevance to real life issues, often helps researchers feel highly motivated about their work.

- **Governmental organisations**. These types of organisations offer a wide variety of jobs for which a thorough understanding of science and technology is often required, but no active research is carried out. National patent offices and regulatory bodies are examples of such institutions. Setting rules and prices for electric power, telecommunication, and so on, requires technical insight in the subject, an understanding of the market economy, and knowledge of how the government regulates these services. Large companies depend heavily on this relationship between government and industry, and their Regulatory Affairs departments welcome PhDs with technological backgrounds and analytical skills.

- **Educational institutions**. Finally, let's not forget to mention the fount of all scientific progress: education. While you must have post-doctoral experience to be able to teach at the university level, colleges and high schools are seeking highly qualified teachers. If you like the idea of shaping young minds, it can be a rewarding experience to educate and inspire the next generation of scientists.

Is Your Final Decision Carved in Stone?

Many PhD students believe that once they have chosen between the academic, for-profit, and not-for-profit sector, they will be obliged to stay on that career path forever. Since these choices are often thought of as being carved in stone, the decision can seem overwhelming and is often postponed. But how final is that decision really? How difficult is it to swing from one branch of the career tree to another? To give you some idea of what this might look like, we'll describe an average situation that could have occurred any time during the last ten years, noting that the flexibility of the job market can play an important role.

Just because you may choose to take a position as a Post-doc, it does not automatically mean that you have to follow an academic career path. Many Post-docs end up in industry or in not-for-profit careers. However, the older you are as a Post-doc, the more difficult it may be to find a job in one of the other branches. For-profit industries are notorious for wanting to hire young and hungry PhDs for their entry-level positions. So if you're sure you ultimately want to pursue a career in the for-profit sector do not hang around working as a Post-doc for more than a year or two.

Going from the for-profit sector to a not-for-profit organisation is a move that can be made at any time. Going from a for-profit industry back to the academic track, however, is more difficult. In practice, only those who work in the research department of a multinational company tend to be successful in making the transition to senior academic positions. Finally, not-for-profit jobs often have the (undeserved!) reputation of being a dead end; once you're in, you'll never get out. But this is certainly not true for all jobs. For instance, once you've become an experienced employee of a patent authority or regulator, you'll also be a very attractive hire for for-profit organisations.

Is There Life Outside the Lab?

What if you've completed your PhD, only to realise that pursuing a career as a lab scientist (whether in academia or industry) is not the right path for you? You may have loved being a PhD student and your work in the lab, but can't envision a future as a professional scientist. So, where do you go from here? Have you just wasted several years of your life pursuing a science PhD? Absolutely not!

If you're someone who relished the independence of working on your own project and enjoyed writing your scientific publications and completing your dissertation, there are plenty of opportunities for you out there that don't involve lab work. For starters, technical or medical writing might be an ideal career for you. In the Pharmaceutical and Biotech industries, plenty of roles need your analytical and science-oriented skills, as well as your keen eye for detail, including Regulatory Affairs, Quality Assurance/Quality Control, and Business Development.

Health Informatics professionals leverage their training in life sciences, information, and IT to manage the enormous amount of data generated by the (bio-)pharmaceutical industry.

Medical Science Liaisons are highly sought after by pharmaceutical companies, medical device companies, and contract research organisations (CROs), among others. Scientific training and an advanced degree (such as a PhD or PharmD) are essential for this role. The job typically involves the following: working with physicians and other medical experts to support clinical trial enrolment, provide protocol training, and ensure that the relevant healthcare professionals possess the most up-to-date information throughout the course of the clinical trial programme.

If the idea of working for a government agency intrigues you, the Food & Drug Administration needs people like you to be reviewers of New Drug Applications or Quality Control Inspectors.

If you're an extrovert and a 'people person', you may enjoy working as a sales representative in the pharmaceutical or biotechnology industries.

Introverts may be more drawn to the writing and editing side of things. Pharmaceutical companies need medical writers for their Scientific Publishing, Regulatory Affairs, and Medical Affairs departments. Academic publishers, such as Elsevier, Wolters-Kluwer, and Springer, to name a few, are always on the lookout for candidates with a strong science background, excellent writing skills, and an eye for detail.

From Searching for Opportunities to Getting the Job

Congratulations! You've figured out which type of job is right for you. Now what? Suppose you have selected a job type based on a decision tree like the one shown in Fig. 22.1, the next move will be figuring out how to get that job. A toolbox of skills to help you in applying for such a position includes the following: writing an effective cover letter and curriculum vitae, learning how to emphasise your strong points and deal with your weaknesses, etc. There are many websites, books, workshops, and courses available on how to acquire the skills you need to get a job; hence, they will not be discussed here. Three specific issues related to PhDs will be discussed, however: (a) When to start your job search, (b) How to use your network (yes you have one!), and (c) How to prepare for a job interview.

When to Start Your Job Search

The key to the numerous steps involved in getting a job is the ability to see things from the employer's perspective. Suppose you are a recruiter for a large company that hires 30 scientists a year. Two candidates come into your office. Candidate Nr. 1 tells you that he got his PhD six months ago. Right after his PhD he took a trip around the world for four months. It was while he was unpacking that he realised he needed a job and that's why he's here. Candidate Nr. 2 says: 'my PhD defence is in two months, after which I plan to take a month's holiday. I am here to find out whether you will have an opening for me three months from now.' Which candidate would you be more interested in hiring? For many hiring managers, proof of a candidate's job planning is essential, and you can demonstrate that skill by timing your application properly and showing your interest ahead time. It should be fairly obvious, unless you're independently wealthy, that you should start your job search well before you obtain your degree.

How to Leverage Your Network

Although the website of your preferred future employer can provide you with a wealth of information, it will lack the richness, essential details, and the kind of important inside information you can only get from a real expert: a former colleague who happens to be working there. Go and talk to him or her. You will get a feel for the company culture, find out what type of job opportunities exist, and get a sense of whether you will like working there once you've been offered a position.

Every career coach will tell you to use your network, but where is this nebulous creature, and do you even have one? As far as you're aware, you don't know anyone who has the type of job you're looking for. However, many PhDs have graduated from your institute in the past five years, and your supervisor or other staff members most likely know where they are now. Contact these former colleagues. They have been in the same position that you're in at the moment and will probably be very happy to discuss the pros and cons of their current employment.

What Type of Skills Do You Possess?

Probably many more than you imagine! Suppose you apply for a position for which a PhD is not mandatory. But the recruiter may likely perceive that by having a PhD, you have certain competitive advantages over other candidates. What are these advantages and how can you demonstrate them? To help you get a better sense of who you are and all that you can do, make a list of all the skills you developed during your doctoral training. Don't forget that your training as a graduate student has taught you much more than just science. You have acquired valuable skills in problem solving, analytical thinking, time management, project management, supervision, giving oral presentations, communication, and teaching, to name just a few.

How to Prepare for an Interview

It goes without saying that being properly dressed and behaving politely are important when you meet your prospective future employer for the first time. There are many websites, blogs, books, and workshops available on how to make a good impression during an interview. Make use of everything that is available to you regarding this important phase in the job-hunting process. At some point, you'll move beyond speaking with the people in personnel and have to start engaging with individuals in the department to which you're applying. Keep in mind that these people are not professional interviewers and meeting new applicants is just one of their tasks. First, they will ask what you have done in the past, what you expect from your new job, and what your long-term plans are. Since you have already prepared the answers to all these questions, you should have nothing to worry about during this part of the interview. But now comes the hard part: you have to become an excellent listener and pay attention to the interviewer when he or she tells you about their job. For the purposes of the interview, accept the premise that they are an expert in the field, and you're completely ignorant, so be careful to act accordingly and avoid coming across as a know-it-all. Sometimes you'll be lucky and happen to know one or two things about the work being described. Most likely this part of the interview, and how well it goes, will play a major role in the final decision of whether to hire you or not.

You can impress the interviewer by being a good listener. Pay attention and try to summarise what the interviewer has been saying (i.e., 'So, if I understand you correctly, you are working on...') and most importantly, ASK QUESTIONS. They know you're not an expert, so feel free to ask if you do not understand things. Most applicants ask too few questions because they're afraid to show their ignorance. You may have to practise asking questions, so try to do that before the interview, either alone or with a friend. Good opportunities for getting accustomed to asking questions can be found at the colloquia you are attending. Force yourself to ask one question at each colloquium. At some point you will get used to it and asking questions during your interview will feel as natural as breathing.

Last, But Not Least: Do You *Want* the Job?

During your job search, you might become so anxious about getting a job, any job, that you may forget to ask yourself whether you really like the position that has been offered to you. Again, stay in the lead. Make sure you have made an active decision based on what you want (e.g., by making a decision tree). Use the interview to get a better feeling for whether you would like to work for that particular department or company. In addition, make a realistic estimate regarding any other positions you might be offered, if you decide to decline a job offer.

In summary, don't wait too long to start thinking about the kind of job you want to pursue after your PhD. Before you write your CV and start sending out applications, you'll need to become familiar with the opportunities available and have a good idea of which job will be the right one for you in terms of furthering your professional goals, both now and in the future.

23

Is Industry Right for You? Opportunities to Explore

Logic will get you from A to B. Imagination will take you everywhere.
—Albert Einstein

In the course of your doctoral studies it may have crossed your mind that your research might have little short-term value—after all, it's mostly curiosity-driven. Join the club. If you're good—or lucky—your research may contribute to society in the long run. If you're ready to work for an organisation that is largely driven by the short-term impact on society, your next step may be to find a job in a for-profit organisation. Although work-for-hire may seem a bit mundane to an academic, this kind of work serves society in the short-, and possibly longterm, in a tangible way, and someone is willing to pay for the service or product you produce.

A wide variety of companies hire PhDs. In this chapter we start by highlighting several aspects of doing research for a company., which might help you decide whether a corporate lab is an appealing destination for you. Next, we discuss opportunities outside a research environment, where you are expected to solve problems, use your analytical skills, explore unknown territory, and learn while you're working; in short, the world of management consultancy.

P. Gosling and B. Noordam, *Mastering Your PhD*,
https://doi.org/10.1007/978-3-031-11417-5_23

Research in a Business-Driven Environment

For better or worse, the science-related corporate world values research differently than that of academia. Industry typically views research as an essential part of its business strategy. For scientists, this means less freedom and, very likely, greater short-term impact. This difference in values generally results in diverse ways of working, and a different set of required skills.

The distance—in time and effort—between a wild idea and a product that can be purchased by a customer is longer than you might imagine, and most potential products get lost (or abandoned) along the way. The steps between wild idea and product vary somewhat with every new product and vary widely between industries. A new gadget for a mobile phone will go through different phases than that of a drug developed by a pharmaceutical company.

Still, almost all products in science-based industries go through research, development, and engineering phases before they enter final testing and production.

Industry's research phase comes closest to the type of work you are most familiar with. Corporate research focuses on the company's long-term product pipeline: the products it hopes to launch over the next 5, 10, or even 15 years. Research activities may not be related to a product the company is already selling or about to launch (this isn't true for pharma companies); rather, the company may be exploring new products or even whole new technologies that may or may not be integrated in a marketed product, eventually.

Not every science-based company has a true research and development department. Even some pharmaceutical companies prefer to bring in promising ideas from outside, by licensing technologies and ideas from other (usually smaller) companies or academic labs.

The first phase in the development of a commercial product is, well… development. Development means transforming the initial concept into a prototype that demonstrates the product's desired functionality. In the development phase, the research ideas are either demonstrated to be feasible for production or abandoned.

Finally it takes a lot of engineering to figure out how to produce a reliable product that has the desired functionality and quality, at reasonable cost. The engineering phase is most likely the furthest away in scope from the type of research you're most familiar with, but the challenges are no less formidable, as are the costs to achieve the desired result in a given time frame.

The early stages of the research part of R&D can be still quite academic; they are, however, largely motivated by the company's strategy and product roadmap. In the design and engineering phase, cost and timing issues are crucial. A good idea—or even a good product—cannot serve a company's business needs if it's too late to market, creates too little additional customer value, or is too expensive to produce.

Some scientists find such constraints disagreeable, considering them to be restrictions on their curiosity-driven work, or a compromise of science for commercial ends. Yet there is an undeniable thrill in knowing that you're working on something that will be actually used, whether it's to make the world a profoundly better place or merely to give pleasure to consumers. Moreover, resources such as people, equipment, and money are much less of a constraint as you get closer towards the final product.

The Difference Between Academic and Corporate Problem Solving

One important difference is that in academia the line between colleagues and competitors is vague, whereas in corporate research it is quite clear: the 1000+ people employed by the same global company work together towards the same ends. Communication is fast and open within the company, while interactions with the outside world are restricted: the process of publishing is carefully timed and controlled in industry so as not to put the company's IP at risk, or to disclose too much of its strategy. In contrast, the academic world is not so starkly divided; today's competitor may well be tomorrow's collaborator, and vice versa. Alliances are always shifting.

In an academic setting, a surprising result can be a breakthrough, providing a new way of looking at a problem. Even if it isn't a breakthrough, you always have the freedom to redefine your problem and keep your science marching forward. In such a setting, serendipity can be an important ally (even though some researchers claim, once the work is done, that a break-through was intentional and anticipated). This academic attitude—being relatively relaxed about the direction of research—has facilitated much of the progress in fundamental science. It has the additional benefit of allowing researchers to plumb the depths of a new scientific insight, hitting a problem hard until it yields its secrets. A process you probably experienced during your PhD!

Corporate researchers rarely have such freedom. In a corporation, research is an expense that is expected to generate revenues in the future. Only a small fraction of conceivable research has profit potential, given a company's expertise and other resources, and only the most promising projects are likely to move forward. Every new project is a roll of the dice, and an unexpected result requires a whole new calculation of risk and potential benefit.

Thoroughness, too, must often be sacrificed: Too bad that problem can't be solved completely; in industry what's needed, especially in the development and engineering phases, is a way forward that is on budget and on schedule, and not necessarily a completely rigorous solution. There may not be time for an elegant, scientifically rigorous approach—but a practical answer to a difficult problem has its own sort of elegance and can be equally satisfying for researchers.

Another consequence of the strong connection between research and the product roadmap is that sometimes projects end prematurely. If your company decides for financial or strategic reasons to abandon a line of activities, research in that area will likely be curtailed. You may be required to abandon your professional passion of the last months or years and embrace something entirely new. It can be frustrating but, in the long run, regular changes of direction can be stimulating and good for your career.

Ways to Explore Whether Corporate Research is Right for You

Companies provide information, via their websites and recruiting brochures, about their (research) activities 'and employment opportunities. But such information tends to be quite general and promotional; it's a start, but really you need something more direct. As a first step, attend career fairs and other career events. Talk to recruiters or whoever else is manning the company's booth. This is an effective way to get a feeling for what type of folks they are and who it is they're looking for.

A better way to investigate is to talk to people within the company with backgrounds similar to yours. Perhaps a former fellow Post-doc or grad student works for the company or knows someone who does. Don't be shy; pick up the phone, make an appointment, even if you know them only remotely. Most likely they will make time to meet you face-to-face and show you the type of work they are doing. Some companies incentivise their employees to make time for these contacts and provide a premium if they help recruit talented former colleagues.

But the best way to figure out whether a particular corporate work environment is right for you is to spend some real time there. Many companies provide opportunities such as fellowship programmes that allow PhD/Master's degree students to experience the corporate work environment.

Not Your Final Destination

Leaving an academic research post can be tough, especially once you have tenure. Moving from a corporate research career back to academia is difficult, though not impossible; only a handful of people make the leap, especially in more applied programmes like engineering. At the managerial level, academia appreciates the corporate experience more. While it is difficult to return to basic science, once you are in the corporate world, it's relatively easy to branch out within a corporate setting.

After several years of R&D experience, many people are eager to take on new challenges. Many make the shift into management, production, sales, customer service, or intellectual property. A corporate lab might be an interesting environment to do research in with a slightly different twist. It can also provide a stepping stone into new kinds of stimulating work later on.

A Career as a Management Consultant

Assisting corporate executives with their toughest decisions may not seem the most obvious career move for someone who has just finished—or is in the process of finishing—a science PhD. But many consulting firms hire PhDs to join multidisciplinary teams to do exactly that, and new PhDs are often thrilled to work in such a novel and exiting environment, where facts and analysis play an important role. As an example of a non-research related continuation of your career after your PhD, we address here a number of aspects of entering the world of management consultants.

Consultancy firms help company managers deal with all kind of issues and problems that arise within their businesses. Of course, companies have internal resources to address their problems. But corporate executives may decide that a certain issue calls for a team of external, independent problem solvers working full time. (See inset for an example.) Typically, consultancy firms send in a small team of consultants to address the issue, supported by partners and expertise from the firm. Usually the team includes a leader responsible for running the daily operation, senior team members with several years of experience, and some younger team members—such as freshly minted PhDs—who are learning on the job.

Solving a corporate problem is not all that different from solving a scientific problem. It requires data, a thorough analysis of the data, and a synthesis of the issues that leads to the best possible solution. Finally, the solution should be reported in such a way that the audience accepts the message and is willing and able to implement it. Those challenges are familiar to most scientists fresh from a PhD programme.

There is one big difference: time. Time is money in the corporate world, particularly for the types of problems that management consultants are usually asked to solve. So it is essential to find the best possible solution within a given time frame, rather than a completely correct 'scientific' answer.

Consultancy Firms Hiring PhDs

Having a PhD is not a prerequisite to joining a consultancy firm, but quite a few management consultants do offer a PhD track. For example, the Boston Consulting Group and McKinsey and Company both have special entry levels for PhDs. Since your PhD research topic is probably of little value to the consultancy firm, you might wonder why they are willing to hire science PhDs with probably irrelevant graduate experience. Firms see problem solving is a key asset that PhDs have. Not just the analytical skills, but also the ability to structure a problem by seeing both the big picture and the details driving it. Being new to the field and approaching problems in an open-minded—though fact driven—way offers something refreshing to an organisation struggling with problems about how to develop and execute its strategy. When joining a consultancy firm, your analytical and quantitative skills are probably just fine, while you may have to catch up on skills and knowledge related to business, finance, and the economy. Firms have the experience that PhDs quickly learn the knowledge and skills they are lacking.

The Steep and Uncertain Career Path as a Management Consultant

Top-tier consulting firms generally have a fast career track; you are expected to move up to the next role within 2–3 years. What if you can't, or don't want to, make the next step up? In that case, most firms would advise you to look for opportunities outside the company.

Is this something to worry about? Probably not. Most former consultants say they learned a lot while on the fast track and received good advice on how to move on in their careers and on what to do next. 'In the long run you are better off learning fast and moving on' when your progress slows, says one seasoned pro.

To learn more about management consultancy, most firms organise business courses or master classes for potential hires. In such a programme, typically lasting a few days, you get the chance to work on a real problem, supervised by consultants. It's an excellent way to gain an appreciation of the thrill of that type of job—or to realise it's just not your cup of tea.

Applying for a job at a management consultancy is not much different from applying for a job anywhere else. Try approaching someone in the company you know, or someone one of your colleagues or friends knows. Follow up with an application letter that states your interest and willingness to work for the company. The initial interviews—which are usually conducted by a recruiter—are likely to be similar to conventional interviews in which you talk about your skills, your career history, and your ambitions, and ask questions about the company.

Your next round of interviews may include working on a case study with one of the company's consultants. You'll receive information about a particular problem and, with help from the interviewer, devise a problem-solving approach and try to crack the problem on the spot. Interviewers are aware that you aren't an expert, so they will focus instead on general skills and the progress you make on the case. Since this approach differs widely from a normal interview, consider doing a practice case upfront. Most company websites provide examples of pertinent case studies. But, as one recruiter suggested, 'the best piece of advice I can give candidates is to get a good night's sleep so that you arrive for the interview well rested and ready to meet the challenges thrown your way'.

Companies Care

Although the years of lifetime employment are over, most companies do care about your career, albeit for sometimes selfish reasons. Academic science still follows the 'tournament model', with all but the most accomplished researchers often feeling as if they're being taken for granted. Once you've survived and earned tenure, your academic freedom will allow you to do whatever you like. But this, in fact, may change little with regard to your current situation; outside your narrow research world, no one truly cares whether you succeed or not. Companies, in contrast, often treat their people—especially their knowledge workers—as their greatest assets. They invest in developing, training, and retaining their staff by nurturing their key employees and hanging onto the staff they need to meet their strategic objectives. This holds true for research-driven companies, consultancy firms, and most other private organisations hiring PhDs.

In summary, although working in a corporate research lab, or as a management consultant, are somewhat unusual entry-level jobs for a newly minted PhD, they have quite a few features in common. Once outside your comfort zone of academic research, you will learn many different ways of working and will have the opportunity to improve your range of skills and knowledge base. You'll also encounter quite a variety of different people and ways of working, with all the pros and cons that go along with a new environment. Continuation on the academic research track is often perceived as a choice for life. Entry-level jobs in a for-profit organisation, however, are often a first step in a career that can branch out in many unexpected directions.

Your First Assignment as a Management Consultant

Because management consultants usually deal with a variety of problems, there are no 'typical' assignments. But here's one example of the type of assignment you might get as a new hire at a consultancy:

The company HighTech is losing market share on its main product, because last year a competitor introduced a superior product. To survive, your client needs to regain its market share by improving the performance/cost ratio of its main product—and its primary moneymaker. This means expanding the company's research and development (R&D) efforts.

In addition, HighTech has a breakthrough technology in the works, but it has to be launched in time for next year's holiday sales. Unfortunately, the new product has major technology-related uncertainties and these, too, require a great deal of input from R&D.

It's up to you and your team to analyse HighTech's current position, evaluate the major technology challenges, consider the options, and decide whether (and how) to pull additional money from the market (loans or stock issuances, for example) to finance these options. And you'd better hurry, because High-Tech is losing money every day and will be bankrupt by next spring if the recovery plan doesn't succeed.

24

Exploring Not-for-Profit Organisations

How wonderful it is that nobody need wait a single moment before starting to improve the world.
—Anne Frank

Perhaps you've decided not to pursue a career in academia, but you're not sure about jumping into an industry job, either. Have you considered the nonprofit sector? If you're passionate about a particular issue, mission-driven, and—in addition to your passion for science—keen to improve the state of humanity and the world, you may be a perfect candidate for a job with a nonprofit organisation. This chapter aims to give you a sense of the breadth and scope of the possibilities open to you, as well as a general sense of what nonprofits have to offer.

Where to Start

Much more than soup kitchens and humanitarian aid, the legion of nonprofit organisations focus on nearly every issue you can think of: environmental protection, climate change, biomedical research, health care, education, international aid, disaster relief, science policy, and science awareness, among countless others. In the United States alone, more than 12 million people—around 10% of the workforce—are currently employed in the nonprofit sector. Think tanks, scientific societies, and foundations are all looking for PhD-level scientists.

© The Author(s), under exclusive license to Springer Nature Switzerland AG 2022
P. Gosling and B. Noordam, *Mastering Your PhD*,
https://doi.org/10.1007/978-3-031-11417-5_24

If you don't wish to stray too far from the lab bench, you can seek research jobs at the many well-known and well-funded nonprofit organisations dedicated to biomedical research; examples include the Howard Hughes Medical Institute (United States), the Fred Hutchinson Cancer Research Center (United States), the Monash Institute of Medical Research (Australia), and the Max Planck Society (Germany).

Numerous smaller nonprofits, such as the Robert Packard Center for ALS Research and the Australian Stem Cell Centre, to name just two, have a specific biomedical focus. Some midsized nonprofit organisations that hire biomedical researchers include SRI International (a nonprofit scientific research institute focused on innovative technologies) and the Institute for OneWorld Health, the world's first nonprofit pharmaceutical company, both based in California. These small and midsized organisations attract excellent researchers who are passionate about their work and committed to research in areas typically neglected by profit-making companies.

If you've decided to pursue a career outside the lab but would like to stay in science, the nonprofit sector has many options for you, including, for example, working in science education for an organisation like the Society for Science and the Public, as an outreach coordinator for the American Association for the Advancement of Science (the organisation that publishes the job search and information website: jobs.sciencecareers.org), or as a programme officer at the Alfred P. Sloan Foundation. PhD scientists in agricultural or environmental sciences may choose to pursue a career at one of the many nonprofits active in the developing world, such as The Nature Conservancy and Earthwatch Institute.

Advantages

There are many issues to consider before starting on a career path in the nonprofit sector, including your personal and professional career goals. Here are a few of the advantages:

- *A wide range of dedicated and like-minded colleagues.* Nonprofits often have their pick of the brightest and most dedicated candidates, many of whom share your values. Staff members typically have a passion for their work and are committed to effecting social change. The result in many cases is an atmosphere of passion, teamwork, and collaboration.

- *Excellent opportunities for professional growth.* Due to the lower staff-to-project ratio in many smaller organisations—and often a flatter organisational structure—you may be assigned several projects and a wide range of tasks, offering you a better-than-average opportunity to strengthen your skill set.
- *Flexibility.* Compared with a corporate enterprise, nonprofits may offer more flexibility in setting and achieving goals, establishing benchmarks, and determining strategies for meeting the organisation's mission.

Disadvantages

Like any sector, nonprofits have some potential downsides:

- *Lower salary.* Most, but not all, nonprofits pay salaries lower than those in industry. This is especially true of advocacy organisations. But there's a wide range across the nonprofit sector, so don't let this particular issue discourage you.
- *Higher employee turnover.* There are many reasons for employee turnover in the nonprofit sector; burnout is high on the list, particularly if the organisation is understaffed, and you are required to multitask. Staff members may leave for better paying jobs, to switch sectors, or to return to school. Often, the smaller nonprofits lack professional development tools aimed at retaining employees.
- *Limited opportunities for career advancement.* At smaller nonprofits, like most small organisations, upper management is very stable, so you might have to switch to another organisation to advance in your career.
- *Structural differences.* If you thrive on hierarchy, discernible targets, and clear deliverables, small and medium-sized nonprofits might not be for you. Organisational clarity may be lacking, as these smaller nonprofits strive to fulfil their missions with limited staffing and resources. In larger and better funded organisations, differences with industry tend to be less pronounced.
- *Fundraising.* Depending on the type of nonprofit, much time will be spent raising funds and writing grants.

One big advantage for many is that nonprofit organisations, whether large or small, are committed to their mission, not to shareholders or to maximising the bottom line. This philosophy, however, may lead to an organisational structure and management style that, for better or worse, can create tensions and obstacles regarding the best way for the organisation to meet its mission and goals.

Is the Nonprofit Sector Right for You?

How will you know if working in the nonprofit sector will be a good fit for you? Understanding yourself and your personal and professional goals is a first step. If you're passionate about a particular issue, and you like the idea of giving back to the community or making the world a better place to live, then a nonprofit organisation may be the perfect place for you to launch your career.

Whether you're considering a job at a particular nonprofit or looking for a nonprofit organisation to work for, you should, as in any job search, thoroughly research the organisation while keeping a few sector-specific issues in mind as you explore your options.

First, look carefully at the organisation's mission statement, which you can almost always find on its website, to decide whether its mission is one you're passionate about (or, at a minimum, one you can believe in). Next, look at the organisation's staff profiles to see if the type of people it employs fits your skills and career ambitions. Finally, take a look at the annual report to see what kind of operating budget the organisation has and how funds are allocated. This information will give you crucial clues about how the organisation goes about achieving its mission, the type of people it hires, and the stability of its finances. Try, if possible, to contact someone who works for the organisation, or a former employee, to get a better sense of its structure, culture, and day-to-day operations.

Once you make it to the interview stage, no matter the size or renown of the organisation, ask some key questions to get a better sense of the organisation's operations and how they treat their staff:

- What is the structure/hierarchy of the organisation? How are decisions made and communicated to the staff?
- Does the organisation encourage teamwork and collaboration, or do staff members work independently on projects?
- What opportunities are there for advancement within the organisation and/or partner organisations?
- What kind of training will I receive? What opportunities exist for professional development?
- What are the organisation's near-term and long-term goals?

- How is the organisation funded and what type of operating budget does it have?

The answers to these questions may help you align your possibly idealistic expectations with how things really work. Even if the organisation is doing excellent work to improve the state of humanity, you can be sure that petty grievances, turf wars, and other aspects of interpersonal friction will be present—just as in any profit-making corporation or bureaucracy-laden government institution.

Finally, ask yourself the following: Am I passionate about the organisation's mission? Is this a place I would like to come to work every day? Do the organisation's goals and objectives fit with my own interests and values? Do I see this as a first step in a career progression of increasing responsibility?

The nonprofit sector isn't for everyone, but for many, particularly those making the transition from academia, the values and culture inherent in the nonprofit world may offer an exciting and rewarding career choice.

Swinging from Branch to Branch on the Career Tree

When starting out in your post-PhD career, the world of work may seem to be rigid and inflexible. PhDs looking back on their career, however, usually notice that they have actually been swinging through the career tree quite a bit. Here is what the members of our team decided to do once they'd obtained their degrees:

> **Saving an Old Master Painting: Epilogue—Diverse Career Tracks with a Touch of Art Restoration**
>
> **Isabel,** ever since she was a child, has been fascinated by paintings, especially the Old Master paintings in which some of the figures are mysterious and barely visible. Having studied chemistry as an undergraduate at university, she still wanted to do something with art. She learned about the scientific aspects of art restoration at a chemistry conference on polymerisation processes. Then she embarked on her PhD programme as discussed elsewhere in this book. As the end of her PhD studies drew closer, Isabel began to realise it was time to start looking for a job. Unfortunately, she feared that the chances of finding one in the field of art restoration were slim, especially for a chemist. What to do? Isabel made an attempt to analyse what she liked most about doing a PhD. After some discussions with friends, Isabel realised that she appreciated in particular two aspects of her PhD programme. First, she enjoyed using

multiple analytical techniques (X-ray, mass spectrometry, IR imaging, etc.) to arrive at scientific conclusions about the nature of the paint pigments. Second, she liked interacting with experts outside the world of natural sciences, who put the scientific results into a broader context. For example, the art historians' perspectives were essential for obtaining a complete picture of the original painting. She liked working with real masterpieces of art, but it was a less essential part of her work. But once she identified her two favourite aspects of the research, no particular job came to mind. One night, she watched a forensic crime series on television. Isabel realised that a job in such a laboratory was actually a perfect match for someone with her ambitions. To Isabel's surprise her professor knew the head of the analytical research department and he arranged for Isabel to meet with him. During that visit Isabel learned that they indeed used a wide range of analytical techniques at the forensic lab, most of which she knew already how to perform. Furthermore, she found out that the interactions with non-scientific experts, such as police inspectors, was an important part of the job, and necessary for solving criminal cases. Well-motivated and prepared, Isabel applied for the job. Five years later, she is head of the analytical services department of the forensic laboratory in her region.

Yousef, after obtaining his PhD in physics, accepted a position in the R&D department of a multinational oil company. After a few years of basic research on oil derivatives, he moved to the consultancy branch of the same multinational to work on green alternatives to fossil fuels. Now he is responsible for in-house advice on novel innovations for reducing the impact of carbon emissions on the environment and for regulatory affairs, along with a lawyer and a marketing consultant. When asked whether working on a PhD was a valuable investment, Yousef always replies: 'Absolutely. In particular, the skills I learned from solving complex problems within an interdisciplinary team turned out to be very valuable.'

Peter, a talented art historian, initially did not consider pursuing an academic career. After a short Post-doc, he joined a leading Information and Communication Technology company. Although he was working on demanding and complex problems, he missed the exploratory nature of basic research in art restoration. At an alumni event, his former supervisor mentioned that there was a starting position available in art history at a nearby university. Peter went to the job interview and suggested to the search committee that they hire two assistant professors rather than one, and that the additional assistant professor have an analytical chemistry background. The university liked the idea and contacted the dean of the natural science department. With some extra money from an art restoration foundation they set up this unique interdisciplinary team. Peter is enjoying his new job, happy to be back in the world of art history and restoration and working again with the chemical side of this challenging field.

25

Lessons Learned

The first step in the acquisition of wisdom is silence, the second listening, the third memory, the fourth practice, the fifth teaching others.
—Solomon Ibn Gabirol

This book describes numerous situations that graduate students typically encounter as they work towards their doctorate. Starting from your very first day in the lab, to the beginning stages of your post-PhD job search, to the day you receive your degree, we've tried to offer you sage advice on how to handle particular situations as they arise. Although individual circumstances are never the same for everyone, we aimed to give you some general guidelines about what we believed would help you make the most out of your years in graduate school. Much of the advice we focused on is of a practical nature and dedicated to the problems typically encountered by most PhD students.

If you glance back through the various chapters, you will notice that we have tried to suggest a general strategy for tackling particular problems. It may seem paradoxical, but similar actions are typically required to solve inherently different problems—no matter what stage of learning and life you happen to be in. And the good news is that many of the skills you learned during your PhD years, regardless of the topic of your doctoral thesis, will be useful in subsequent stages of your career.

In this final chapter we summarise, in general terms, the key lessons we hope you have learned. Perhaps the two most important, and the ones that run through all the chapters like a common thread, are ***proper planning*** and ***good communication***. If you've recognised the importance of these two skills

© The Author(s), under exclusive license to Springer Nature Switzerland AG 2022
P. Gosling and B. Noordam, *Mastering Your PhD*, https://doi.org/10.1007/978-3-031-11417-5_25

and have managed to put them to good effect, you have probably come to the end of your graduate study with a sense of accomplishment, as well as pride in yourself for having successfully survived the course. In fact, the strategies you have acquired in learning how to master your PhD will be useful in every job you have, both now and in the future. To repeat a statement that we included in the introduction: former PhD students claim that the communication, planning, and problem-solving skills they acquired during their PhD research are as useful to them, if not more, as the actual content that went into their thesis.

Planning is Essential

Scientists tend to be sceptical by nature. It goes with the territory. Thus, many of them claim that you simply cannot plan science. Research, after all, has a life and rhythm of its own. So, some of you may have felt that trying to plan your time in the lab was a wasted effort. Indeed, you cannot plan the outcome of your scientific efforts. But we still believe in the importance of proper planning (and this includes time management), and that good planning will maximise your chances of getting the most out of your time in the lab. There's more to grad school than research. All kinds of 'fringe activities' will take up your time and pull you away from the lab bench. Attending conferences, discussing work, preparing presentations, reading the literature, searching the internet, and managing your e-mail, are all essential activities. With proper planning you can optimise the results of those fringe activities so that you spend as little time as possible on things that are essential but take time away from your thesis research.

Every now and then it helps to take a step back and look at your world with an impassioned eye. If you can honestly assess the productiveness of everything you did, you may realise that you could probably have skipped a good fraction of those activities (per the 80/20 rule). Truly wise planners know how to stay one step ahead of their problems. By reflecting on what went wrong in the past and being honest about what might go wrong in the future, you will be able to take appropriate measures that will save you a lot of time. Therefore, we have repeatedly suggested that you identify the potential hurdles you might have to leap over, as well as the pitfalls you need to avoid, for instance, during your monthly progress review. Once you have identified potential problems you can consider alternative approaches for obtaining the result you want.

People with a more chaotic approach to projects (i.e., Perceivers in MBTI terminology), often find that things do eventually work out the way they want in the end—despite their lack of good planning. This may work only if your actions, and yours alone, are required to obtain the result you're seeking. But doing a PhD always requires the help of others, to a greater or lesser extent. Thus, teamwork is essential, and getting people to work in harmony takes time and effort. Very few people will start marching to the tune of your drum, even if you think that's the best way to proceed. You have limited power to change the behaviour of others, so in order to ensure that they deliver on their promises, you will have to plan carefully and be realistic about lead times for all the things (and the list is long) that are not under your control. Of course, for a plan to be successful, you will need to communicate the planning to others. This brings us to the second valuable skill we hope you have practiced and refined along the way.

Communication Creates Harmony

Even if it's just your name that appears on the spine of your printed and bound thesis, remember this: you are part of a team. You might not find that your teammates are active or visible at all times, but they are there. From librarians to roommates, from undergraduate students to your supervising professor, they have all contributed to your research in one way or another. Share your good news with others (celebrate your success) and ask for help and advice if you are making less progress than you expected. Be able to admit it if you don't know the answer to something. Ask questions. Listen to the answers. If you have an open attitude and make clear what you expect from others and what they can expect from you, your years as a graduate student will be more productive, not to mention much more pleasant.

Misunderstandings that arise from a lack of communication are the source of many conflicts and much unhappiness. Solving conflicts is essential to moving forward with your team. The first step in solving interpersonal problems is communication. A key feature of good communication is active listening, a skill that is also of great value during job interviews and meetings. As you make steady progress in your project, you might forget to communicate properly on a regular basis. When work needs to be done, it may seem to you that talking, and especially listening, is a waste of time. But it's essential to keep communicating your progress, as others might have interesting suggestions to speed up your research or to redirect your experiments, so that the scientific value of the results is greater in scope. A plan not communicated with your team is bound to fail. The converse is also true: communication that is not planned loses much of its value.

Of course, you already know that you have to prepare in advance your presentations at conferences and group meetings, but your yearly evaluations, as well as any other meetings you have with your thesis advisor, will have a bigger impact if you are thoroughly prepared beforehand. Think about how important something as trivial as a proper subject line is when you write an e-mail. Words carry a great deal of impact. A well thought out subject heading will have a much better chance of being acted upon by the recipient. So it is with talking and listening to the members of your team.

Some Final Thoughts

The high-level research and prodigious effort that go into earning a PhD require hard work, dedication, and the ability to recover quickly from setbacks. At times, it may seem like a lot of work, and no play whatsoever, but there are also many upsides to tackling a challenging scientific project and seeing it through to the end. Enjoy and celebrate your successes—such as the occasion when you were the first person to obtain novel data about a particular topic and drew important conclusions from it. Doing science in a research institute is a job, certainly, but it's a special one. Some even say that it's a calling. You're surrounded by young, hard-working, and ambitious people. But don't forget that there are also many opportunities to have fun with your fellow students. You're all in this together, so don't be afraid to inject levity and humour into the day-to-day seriousness of your work. A PhD programme is much more than getting results and writing a thesis. It's a period in your professional life in which you acquire many new skills and make professional and personal contacts that will last a lifetime. Allowing yourself the opportunity to master those skills, as well as have fun in the process, is the sagest advice we can offer.

Acknowledgements

The authors would like to thank Prof. Dr. Katrien Keune, Head of Science, Rijksmuseum Amsterdam, for her help with the technical details in the 'Saving an Old Master Painting' narrative; Herman Roozen of Comic House Oosterbeek for the illustrations that grace the text; Belle Jansen, communications expert, University of Amsterdam, for the editorial contributions to Chapter 15 (Remote Collaboration); and Marc Noordam, PhD student at Delft University, for his insights on life as a PhD student during the Covid-19 pandemic. Some of the material originally included in the second edition (and retained in the third edition) was published as part of a monthly column on the ScienceCareers.org website. We thank James Austin, the magazine's editor, for inspiring us to expand our reach in providing additional insights and advice to PhD students.

About the Authors

Dr. Patricia Gosling obtained her PhD in Bioinorganic Chemistry from the University of Nijmegen, the Netherlands. Dr. Gosling has spent most of her career in the pharmaceutical and medical communications industries and has lived and worked in the Netherlands, Germany, Malaysia, and Switzerland. After twenty-plus years in pharma/biotech as a regulatory and clinical medical writer and document specialist, she currently works as a freelance academic editor, with a focus on the natural sciences.

Prof. Dr. Bart Noordam obtained his PhD in Physics from the University of Amsterdam, the Netherlands. Dr. Noordam is Senior Vice President Strategy at ASML Netherlands BV, an innovation leader in the semiconductor industry. Previously, he was Dean of the Faculty of Science at the University of Amsterdam. As Professor of Physics he supervised PhD students over the span of a decade. He has also worked as Director of a National Research Institute in Amsterdam, as a strategy consultant for McKinsey & Company, and is the founder of a Regional Audit Organisation for provinces in the Netherlands.

Printed in the United States
by Baker & Taylor Publisher Services